BASIC PHYSICS OF STELLAR ATMOSPHERES

Astronomy and Astrophysics Series

Editor: A. G. Pacholczyk

INTERMEDIATE SHORT TEXTS IN ASTROPHYSICS

BASIC PHYSICS OF STELLAR ATMOSPHERES
T.L. Swihart

PHYSICS OF STELLAR INTERIORS
T.L. Swihart

In preparation:

INTRODUCTORY THEORETICAL ASTROPHYSICS
Ray J. Weymann, Editor

RELATIVISTIC ASTROPHYSICS
W. John Cocke

RADIO SOURCES
A.G. Pacholczyk

BASIC PHYSICS OF
STELLAR ATMOSPHERES

Thomas L. Swihart

Steward Observatory
The University of Arizona

Second Printing

PACHART PUBLISHING HOUSE

Tucson

Printed in the United States of America

Library of Congress Catalog Card Number: 72 - 180256
International Standard Book Number: 0-912918-04-7

Copy Editor: Neil D. Lubart
Art Director: Mary Jane Fogarty

Pachart Publishing House
3920 East Fort Lowell Road
Post Office Box 6721, Tucson, Arizona 85716

Pachart Japan
Tanimichi Sugita, Representative
Co-op Olympia, Suite 200
35-3 Jingumae 6-chome
Shibuya-ku, Tokyo
150 Japan

CONTENTS

PREFACE

This book is intended as a short text on the theory of stellar atmospheres for the non-specialist. The reader is expected to be familiar with mathematics through differential equations and to have a fairly solid undergraduate physics background.

As a text, the book develops the subject in logical steps, the relevance of new topics is discussed at the time they are introduced, and emphasis is given to the relation each part has to the whole. Although mathematical details are given in a few places, this is essentially a physical introduction to the subject.

The book has been deliberately kept short. There are many details which are essential to carrying out research in the field but which are not required for an understanding of the subject. Most of these details have been omitted, as I feel that they would interfere with the main purpose of the book, which is to give a good basic understanding to the qualified reader. Some will argue that I have misplaced the emphasis on certain special topics, such as non-LTE, model construction, abundance analyses, etc., but the text is what I consider to be a balanced presentation of the different parts of the subject.

In the first chapter, the basic quantities of radiation transfer are introduced and the equation of transfer is derived. In the next two chapters, various aspects of the problem of the energy balance in a plane semi-infinite medium having a significant radiative flux are considered. Chapter 2 is concerned primarily with the much simplified but still important case in which the absorption is independent of frequency, while Chapter 3 looks at more physically realistic theoretical models of stellar atmospheres. The transport of line radiation is the subject of the final chapter. The text material is extended by a set of problems given in Appendix 2. A list of general references is given at the beginning of the text, and further references are given at the appropriate places in the text.

I wish to express thanks to E.H. Avrett, N.D. Lubart, and A.G. Pacholczyk for comments and support during the preparation of this work.

THOMAS L. SWIHART

The University of Arizona
Tucson, Arizona
August, 1971

BASIC PHYSICS OF STELLAR ATMOSPHERES

GENERAL REFERENCES

Aller, L.H. *The Atmospheres of the Sun and Stars*, 2nd ed., Ronald, 1963.
Avrett, E.A., Gingerich, O.J., and Whitney, C.A. (eds.) *First Harvard-Smithsonian Conference on Stellar Atmospheres*, S.A.O. Special Report No. 167, 1964.
__________. *The Formation of Spectrum Lines* (2nd Harvard-Smithsonian Conference on Stellar Atmospheres), S.A.O. Special Report No. 174, 1965.
Böhm, K.H. "Basic Theory of Line Formation", in *Stellar Atmospheres* (J.L. Greenstein ed.), U. of Chicago, 1960.
Chandrasekhar, S. *Radiative Transfer*, Oxford, 1950.
Gingerich, O.J. (ed.) *Theory and Observation of Normal Stellar Atmospheres* (3rd Harvard-Smithsonian Conference on Stellar Atmospheres), M.I.T., 1969.
Jefferies, J.T. *Spectral Line Formation*, Blaisdell, 1968.
Mihalas, D. *Stellar Atmospheres*, Freeman, 1970.
Münch, G. "Theory of Model Stellar Atmospheres", in *Stellar Atmospheres* (J.L. Greenstein ed.), U. of Chicago, 1960.
Pecker, J.C. "Model Atmospheres", in *Ann. Rev. Ast. Astrophys.* **3**, 135, 1965.
Unsöld, A. *Physik der Sternatmosphären*, 2nd ed., Springer, 1955.

1

THE EQUATION OF TRANSFER

1. Intensity and Derived Quantities

The basic unit of radiative energy transport is the intensity. Consider an element of area dA whose normal is along n (see Figure 1). From any point on this surface construct an element of solid angle $d\omega$ around a direction making the angle θ with n. If this is done for all points on dA, keeping the given direction fixed, the envelope of these constructions will be a truncated cone. The cross sectional area of the cone normal to its axis is $\cos\theta\ dA$. The radiation energy which crosses dA in the time interval dt in the given direction and confined to that solid angle is proportional to the projected area, to the interval of time, and to the solid angle. The factor of proportionality is the intensity I:

$$dE = I \cos\theta\, dA d\omega dt. \tag{1.1}$$

Thus intensity is energy per area per solid angle per time, and it is a function of position, direction, and time. It is also a function of the wavelength or frequency of the radiation. Thus, there is a monochromatic intensity I_ν defined such that $I_\nu d\nu$ is the intensity due to all waves with frequencies between ν and $\nu + d\nu$. There is a similar definition of the monochromatic intensity per unit wavelength interval I_λ. The total or integrated intensity is then

$$I = \int_0^\infty I_\nu d\nu = \int_0^\infty I_\lambda d\lambda. \tag{1.2}$$

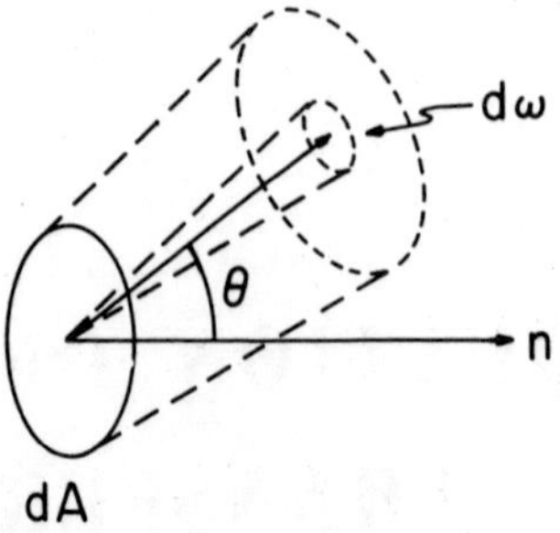

Fig. 1. Defining intensity of radiation.

Frequency dependence is not important in the present section, and it will not be explicitly indicated. The relations of this first section will be valid both for monochromatic and for integrated quantities.

An important quantity derived from the intensity is the mean intensity J. This is the intensity averaged over all directions. Since the total solid angle around a point is 4π steradians, one has

$$J = \frac{1}{4\pi}\int I \; d\omega = \frac{1}{4\pi}\int_0^{2\pi}\int_0^{\pi} I(\theta,\phi) \sin\theta \; d\theta d\phi. \tag{1.3}$$

In the second expression above θ and ϕ are spherical angles.

Another quantity derived from the intensity is the flux. This is designated in the astrophysical literature by πF, so that F alone is the physical flux divided by π. This convention is unfortunate, but it is so well established that it will be adhered to in the present work. Sometimes the quantity $H = F/4 = \text{flux}/4\pi$ is used.

Let $\mathbf{s}(\theta,\phi)$ be a vector of unit length in the direction specified by the spherical angles θ and ϕ. Then the flux at any point is the vector given by the following integral over the solid angle:

$$\pi\mathbf{F} = \int I(\theta,\phi) \; \mathbf{s}(\theta,\phi) \; d\omega. \tag{1.4}$$

If $\mathbf{n}$ is a unit vector in a fixed direction, the component of the flux along $\mathbf{n}$ is found by taking the scalar product with the above:

$$\pi F(\mathbf{n}) = \pi\mathbf{F}\cdot\mathbf{n} = \int I \; \mathbf{n}\cdot\mathbf{s} \; d\omega. \tag{1.5}$$

By choosing the direction of $\mathbf{n}$ to define $\theta = 0$, one requires that $\mathbf{n}\cdot\mathbf{s} = \cos\theta$. Thus equation (1.5) becomes

$$\pi F(\mathbf{n}) = \int I \; \cos\theta \; d\omega \, . \tag{1.6}$$

By comparing equations (1.1) and (1.6) one sees that the component of the flux in a given direction is simply the net energy per unit area and per unit time crossing a surface normal to that direction. Unless otherwise stated, the term flux will hereafter refer to the magnitude of the vector flux. It is seen that the vector flux is just the Poynting vector $(c/4\pi)\; \mathbf{E} \times \mathbf{H}$, where $\mathbf{E}$ and $\mathbf{H}$ are the electric and magnetic fields.

If one makes use of the substitution $\mu = \cos\theta$, one can write the flux as follows:

$$\pi F = \int_0^{2\pi}\int_0^{\pi} I(\theta,\phi) \cos\theta \sin\theta \, d\theta d\phi ,$$

$$= \int_0^{2\pi}\int_{-1}^{+1} I(\mu,\phi)\mu \, d\mu \, d\phi,$$

$$= \int_0^{2\pi} d\phi \left[\int_0^{+1} I(\mu,\phi)\mu \, d\mu - \int_0^{+1} I(-\mu,\phi)\mu \, d\mu \right]. \tag{1.7}$$

The first term on the right side of equation (1.7) is due to radiation being propagated in the outward or positive hemisphere ($\mu > 0$), while the second term represents radiation traveling in the negative ($\mu < 0$) direction. If these partial fluxes into the respective hemispheres are denoted by πF^+ and πF^-, then equation (1.7) becomes

$$F = F^+ - F^- \tag{1.8}$$

The total or net flux is thus the excess of that in the positive direction over that in the negative direction. In an isotropic radiation field the latter two quantities are equal and the net flux is zero: there is no net transport of energy in any direction.

Equation (1.1) gives the energy in a given pencil of radiation, i.e., the energy crossing a given surface element in time interval dt, in a given direction, and confined within a given solid angle. This energy is contained in the volume $\cos\theta \, dA \, c \, dt$, and so the radiation in that particular solid angle contributes an amount $I \, d\omega/c$ to the energy per unit volume at the point considered. The total energy density u is

$$u = \frac{1}{c}\int I \, d\omega = \frac{4\pi}{c} J. \tag{1.9}$$

The energy density is thus an equivalent quantity to the mean intensity.

Radiation of energy dE carries momentum dE/c in the direction of propagation. Thus, equation (1.1) indicates that the given pencil of radiation contributes $I \cos^2\theta \, dA \, d\omega \, dt/c$ to the normal component of momentum carried across dA in interval dt. But the normal component of momentum carried across a surface per unit area and time is the scalar pressure at the given point. The radiation pressure is then

$$P_r = \frac{1}{c}\int I \cos^2\theta \, d\omega. \tag{1.10}$$

It should be emphasized that the radiation quantities considered above are defined for any point, not for an extended surface or volume. The flux is the ratio of energy to area, taken in the limit as the area goes to zero. The other quantities are defined in a similar way.

The intensity of radiation is not affected by the geometry of the region of interest. To see this, consider two elements of area dA and dA' whose normals make the angles θ and θ' to the line connecting them. According to equation (1.1), the energy per unit time which crosses dA and which later crosses dA' is $I \cos\theta \, dA \, d\omega$, where I is the intensity at dA and where $d\omega$ is the solid angle of dA' as seen from dA. A solid angle is the projected area divided by distance, so that $d\omega = dA'\cos\theta'/s^2$, where s is the distance between the surfaces. The rate of energy flow from dA to dA' is then

$$\frac{dE}{dt} = \frac{1}{s^2} I \cos\theta \cos\theta' \, dA dA'. \tag{1.11}$$

Consider now the energy per second which crosses dA' and which had previously crossed dA. This will be $I'\cos\theta' dA' d\omega'$, where I' is the intensity of the beam of radiation at dA' and where $d\omega'$ is the solid angle of dA as seen from dA'. Thus

$$\frac{dE'}{dt} = \frac{1}{s^2} I'\cos\theta'\cos\theta \, dA'dA. \tag{1.12}$$

If there is matter between the two surfaces which can alter the energy flow through absorption or emission, then the expressions (1.11) and (1.12) will differ; otherwise, they will be the same and $I = I'$. The intensity along a given pencil of radiation is altered only by physical sources or sinks along the path, but not by geometry. The intensity of starlight as seen from the Earth is the same as that at the surface of the star, except for the amounts blocked out by the interstellar matter between the star and the Earth and by the Earth's atmosphere. The difference is that, near a star, the radiation is coming from a much larger solid angle than the case at a great distance from the star.

As indicated in equation (1.6), the flux due to a source is the integral of $I \cos\theta$ over the solid angle of the source. Since the solid angle of any object varies as the inverse square of its distance, it follows that the flux from a source falls off with distance in the same way. If there is absorbing material between the source and the observer, the flux will decrease even more rapidly with distance.

Consider a spherical star of radius R, observed from the distance r. The measured flux is

$$\pi F = \int_0^{2\pi}\int_0^{\theta_0} I(\theta,\phi) \cos\theta \sin\theta \, d\theta d\phi. \tag{1.13}$$

Here, $\theta_0 = \sin^{-1}(R/r)$ is the angular radius of the star. If θ' is the angle at the surface of the star between the normal to the surface and the direction to the observer, as illustrated in Figure 1.2, then for neighboring points on the stellar surface

$$r^2\cos\theta \sin\theta \, d\theta = R^2\cos\theta' \sin\theta' d\theta'. \tag{1.14}$$

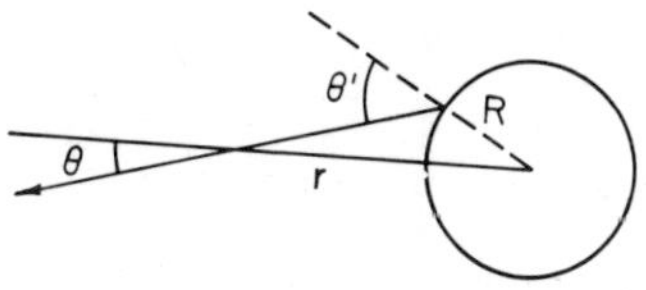

Fig. 2. Geometry of measurement of stellar flux.

Also, as θ varies from zero to θ_0, θ' goes through the range 0 to $\pi/2$. Thus equation (1.13) gives

$$\pi F = \frac{R^2}{r^2} \int_0^{2\pi} \int_0^{\pi/2} I(\theta',\phi) \cos\theta' \sin\theta' \, d\theta' d\phi . \tag{1.15}$$

Since the integrals depend only on the properties of the star in equation (1.15), the flux is seen to vary as r^{-2}. In the special case in which I does not depend upon θ' and ϕ, equation (1.15) reduces to

$$F = \frac{IR^2}{r^2} . \tag{1.16}$$

2. The Absorption Coefficient

This section is concerned with a description of how matter interacts with radiation to subtract energy from the radiation stream. Since most of the relations will be valid for monochromatic quantities but not integrated ones, the frequency subscript will be explicitly indicated. While I_ν will generally be used, I_λ could just as easily be substituted for it.

The rate at which matter causes energy to be lost from a beam of radiation is described in terms of an absorption coefficient. A pencil of radiation of intensity I_ν will lose by absorption the amount dI_ν on traveling the distance ds. The absorption coefficient is defined by

$$dI_\nu(\text{abs}) = I_\nu k ds = I_\nu \kappa \rho ds , \tag{2.1}$$

where k is the volume absorption coefficient of dimensions length^{-1}, κ is the mass absorption coefficient of dimensions length^2 mass^{-1}, and ρ is the mass density of the matter. k and κ are alternate quantities to describe the same thing. In practice one uses the form of the absorption coefficient that is most convenient, and this may be either the mass or the volume coefficient. Both will be used in the present work.

Most matter will absorb some frequencies much more readily than others, so k can be expected to depend strongly on frequency. For this reason k is commonly written with a frequency subscript; however, this is not necessary and it can be confusing: I_ν does not have the subscript simply because it depends upon frequency (it also depends on position and time and direction), but because it is a measure per unit frequency interval. k does not have this measure, and the subscript will not be used here.

Absorption as described above includes any process in which a photon is removed from a pencil of radiation. If a photon is deflected into a new direction without being otherwise changed, the process is called scattering. Non-scattering refers to processes in which the photon is changed into a different form of energy, perhaps to reappear later as another photon. Absorption includes both scattering and non-scattering.

Let $W(s)$ be the probability that a photon will travel the distance s without being absorbed. Then $W(s_1 + s_2) = W(s_1)\, W(s_2)$. With $s_1 = s$ and $s_2 = ds$, one has, using a first order Taylor expansion,

$$W(s + ds) = W(s)\, W(ds) = W(s) + \frac{dW}{ds}\, ds. \tag{2.2}$$

According to equation (2.1), $k\, ds$ is the probability that absorption will take place in ds, so $(1 - k\, ds)$ is the probability that it will not, which is $W(ds)$. If this is put into equation (2.2), and if the integration is carried out under the assumption that k does not change with position, the result is

$$W(s) = e^{-ks}. \tag{2.3}$$

Finally, let $P(s)\, ds$ be the probability that absorption will take place between s and $s + ds$. Then $P(s)\, ds = W(s)[1 - W(ds)]$, or

$$P(s)\, ds = e^{-ks} k\, ds. \tag{2.4}$$

The mean free path L is the average distance a photon travels before being absorbed. From equation (2.4) one finds

$$L = \int_0^\infty s\, P(s)\, ds = \frac{1}{k}. \tag{2.5}$$

Thus the volume absorption coefficient k is the reciprocal of the photon mean free path. If k does vary with position, the integrals leading to expressions (2.3) and (2.4) cannot be carried out, but equation (2.5) still defines the local mean free path.

k and κ are macroscopic coefficients in that they describe bulk properties of matter. Absorption can also be described on the microscopic scale. Suppose that each absorbing particle is represented as a sphere of radius R, so that $a = \pi R^2$ is the effective cross sectional area which the particle exposes to the photons. If there are N absorbers per unit volume, a column of area A and length ds contains $NA\, ds$ of the absorbers, and the total cross section they will show to the photons is $aNA\, ds$. Thus $Na\, ds$ is the probability a photon will be absorbed on traveling the distance ds. Comparing this with equation (2.1) one has

$$k = Na. \tag{2.6}$$

Thus k is the total cross section of the particles contained in unit volume, and hence the name volume absorption coefficient. Similarly, κ is the total cross section per unit mass of the medium.

It is apparent that the volume absorption coefficients are additive, at least as long as the particles are far enough apart that they can act independently. If there are different kinds of particles in a mixture, having coefficients k_1, k_2, etc., then the volume coefficient of the mixture is $k = k_1 + k_2 + \ldots$. The same is not true of the mass absorption coefficients, unless the individual coefficients are given per unit mass of the mixture.

3. The Emission Coefficient

The volume emission coefficient j_ν is defined such that the energy given off per unit time by volume element dV into solid angle $d\omega$ and within the frequency range $d\nu$ is given by

$$\frac{dE_\nu}{dt} = j_\nu \, d\nu dV d\omega. \tag{3.1}$$

The frequency subscript is used here since both j_ν and E_ν are measured per unit frequency interval. As with the absorption coefficient, the emission can be expressed per unit volume or per unit mass. If ε_ν is the mass emission coefficient, then $j_\nu \, dV = \varepsilon_\nu \, dm$, or $j_\nu = \rho\varepsilon_\nu$.

Just as the absorption coefficient must include all effects which subtract energy from a given pencil of radiation, the emission coefficient must include all effects which add energy to the beam. This includes radiation which is simply scattered into the given beam from other directions. The scattered energy must be added to whatever other forms of energy are converted to radiation and emitted by the volume element.

Consider the volume element dV as being the cylinder of cross section dA and length dx. According to equation (3.1), the contribution of dV to the energy per second into $d\nu$ and $d\omega$ is $j_\nu \, dAdxd\nu d\omega$. If ds is the length of dV along the direction of propagation, $dx = ds \cos\theta$, where θ is the angle between the direction of propagation and the axis of dV. If this is compared with equation (1.1), it is seen that the intensity in the given beam is increased by the amount

$$dI_\nu(\text{em}) = j_\nu \, ds. \tag{3.2}$$

The contribution of any small region to the intensity is the volume emission coefficient times the linear size of the region in the direction of propagation.

4. The Equation of Transfer

When radiation propagates a distance ds, the intensity will be increased by the emissions and decreased by the absorptions along the path. Combining equations (2.1) and (3.2), one has

$$\frac{dI_\nu}{ds} = j_\nu - kI_\nu . \tag{4.1}$$

It is understood that ds is measured along the direction of propagation. As one follows a given pencil of radiation, the intensity can vary through its dependence upon position and time. The dependence on direction is not involved since the direction is held fixed for a given pencil of radiation. Then dI_ν is equal to $(\partial I_\nu/\partial s)ds + (\partial I_\nu/\partial t)\,dt$. But ds is the distance radiation travels in time dt, so $dt = ds/c$. In general, therefore, equation (4.1) becomes

$$\frac{1}{c}\frac{\partial I_\nu}{\partial t} + \frac{\partial I_\nu}{\partial s} = j_\nu - kI_\nu . \tag{4.2}$$

It is only rarely that the explicit time dependence of the intensity need be taken into account. This does not mean that time plays no role, but it does mean that the physical quantities generally do not vary appreciably in the time it takes radiation to cross the system of interest. In this case one can set $\partial I_\nu/\partial t = 0$ and any time dependence is brought in implicitly through the absorption and emission coefficients. The derivative with respect to position becomes total.

ds can be related to any desired coordinate system. If q_i are the coordinates of interest, then

$$\frac{\partial}{\partial s} = \sum_i \frac{\partial q_i}{\partial s}\frac{\partial}{\partial q_i} . \tag{4.3}$$

In a cartesian system, $dx = l\,ds$, $dy = m\,ds$, $dz = n\,ds$, where (l, m, n) are the direction cosines of the propagation direction. Then

$$\frac{\partial I_\nu}{\partial s} = \left(l\frac{\partial}{\partial x} + m\frac{\partial}{\partial y} + n\frac{\partial}{\partial z}\right) I_\nu . \tag{4.4}$$

In a system with spherical symmetry, let r be the distance from the center and θ the angle measured from the radial direction. Then $dr = \cos\theta\,ds$, and $r\,d\theta = -\sin\theta\,ds$, so

$$\frac{\partial I_\nu}{\partial s} = \left(\cos\theta\frac{\partial}{\partial r} - \frac{\sin\theta}{r}\frac{\partial}{\partial \theta}\right) I_\nu . \tag{4.5}$$

In terms of $\mu = \cos\theta$, this is

$$\frac{\partial I_\nu}{\partial s} = \left(\mu\frac{\partial}{\partial r} + \frac{1-\mu^2}{r}\frac{\partial}{\partial \mu}\right) I_\nu . \tag{4.6}$$

In a similar way one can write the equation of transfer in any coordinate system. Of course one will use that system which takes greatest advantage of whatever symmetry the problem of interest possesses.

If equation (4.1) is multiplied by the factor $\exp(\int_0^s k(s')\,ds')$, the two terms involving intensity combine to make a perfect derivative. The result is then easily integrated.

$$I_\nu(s_2) = I_\nu(s_1)\exp\left(-\int_{s_1}^{s_2} k\,ds'\right) + \int_{s_1}^{s_2} j_\nu \exp\left(-\int_{s}^{s_2} k\,ds'\right) ds. \tag{4.7}$$

This equation gives the intensity at a given point in terms of properties of the same pencil of radiation at previous points along its path of propagation. Thus the intensity at point s_2 equals the intensity at any previous point s_1, attenuated by the exponential factor which represents the absorption between s_1 and s_2; to this is added all of the emission between the same points, again attenuated by the absorption between the integration point and s_2. If there is no absorption, the intensity increase is simply the sum of all emissions over the path length, as equation (3.2) indicates. If there is no emission, $I_\nu(s_2)$ is the incident intensity cut down by the exponential absorption, consistent with equation (2.1).

If the explicit time dependence is kept, so that the appropriate equation of transfer is (4.2) instead of equation (4.1), then the solution is still of the form of equation (4.7). The only difference is that each quantity in the expression is to be evaluated at its retarded time argument, that is, at the light travel time interval in the past.

A simple case is the solution for a homogeneous medium in which the absorption and emission coefficients do not depend on position. Then equation (4.7) is seen to reduce to:

$$I_\nu(s_2) = I_\nu(s_1)\, e^{-k(s_2-s_1)} + \frac{j_\nu}{k}\left[1 - e^{-k(s_2-s_1)}\right]. \tag{4.8}$$

$k(s_2 - s_1)$ is a dimensionless quantity known as the optical distance or thickness between s_1 and s_2. The two extreme cases of very large and very small optical thickness are of particular interest. One has

$$I_\nu(s_2) = I_\nu(s_1) + j_\nu(s_2 - s_1) \qquad \text{(optically thin)}$$

$$I_\nu(s_2) = \frac{j_\nu}{k} \qquad \text{(optically thick)} \tag{4.9}$$

In the optically thin case, one again has simply equation (3.2). In the thick case, the more distant parts of the source are not important because of the large absorption between them and the point in question. Only points within a few mean free paths give a significant contribution. In fact equations (2.5) and (4.9) show that an optically thick source is equivalent to a thin one whose size is one mean free path.

5. The Source Function

The ratio of the emission to the absorption coefficient is a very important quantity in radiation transfer theory called the source function S_ν:

$$S_\nu = \frac{j_\nu}{k} = \frac{\varepsilon_\nu}{\kappa} . \tag{5.1}$$

Its importance is indicated by the fact that one frequently works in terms of S_ν instead of the emission coefficient. Part of the reason for this is that it is often convenient to use optical depth τ rather than linear depth s as the independent parameter. For instance, the contribution of emission to the intensity is $j_\nu ds = S_\nu d\tau$, where $d\tau = k ds$, so the source function fits in well when optical depth is used. A more important reason for the use of the source function is that it is generally far less sensitive to the detailed properties of the medium than is the emission coefficient. In fact for a range of circumstances wide enough to be very useful in practice, S_ν is a universal function of temperature and frequency, and independent of all other parameters, including the composition of the medium.

Suppose that there are several different processes taking place, each with its own absorption coefficient k_i and its own emission coefficient $j_\nu(i)$. Then, since these coefficients are additive, the total quantities are $k = \Sigma k_i$ and $j_\nu = \Sigma j_\nu(i)$. The total source function is then

$$S_\nu = \frac{j_\nu}{k} = \frac{1}{k}\sum_i j_\nu(i) = \frac{1}{k}\sum_i k_i S_\nu(i) . \tag{5.2}$$

The total source function is the sum of the individual ones, weighted by the individual absorption coefficients, where $S_\nu(i) = j_\nu(i)/k_i$.

The following discussion is intended primarily for continuum processes. For bound-bound transitions, which are considered in Chapter 4, the emphasis is somewhat different, although much of what is stated here will also have application to the lines.

It is convenient to divide absorption and emission into scattering and non-scattering parts, indicated by sc and ns as arguments. A scattering process is one in which the photon essentially maintains its identity through the absorption and emission. The direction and perhaps the frequency are changed, but there is a one to one correlation between the photons before and after scattering. For non-scattering, the absorbed photon is lost to the radiation field at least temporarily, and there is only a statistical correlation between the photons being absorbed and those emitted. The distinction is important because the scattering part of the source function depends directly upon the radiation field, which is the main unknown of the problem. For non-scattering, however, the source function is much less directly connected with the radiation field. Thus the method of attacking a problem may vary considerably, depending on which types of processes dominate. From equation (5.2) one can write

$$S_\nu = \frac{k(sc)}{k} S_\nu(sc) + \frac{k(ns)}{k} S_\nu(ns). \tag{5.3}$$

The scattering part of the source function will be considered first, followed by a discussion of the non-scattering part.

According to equation (2.1), the intensity of frequency ν' lost by scattering over the path ds is $I_{\nu'}k(sc)ds$. Then the energy lost in time dt and frequency interval $d\nu'$ from the pencil of radiation defined by the area dA and solid angle $d\omega'$ is

$$dE_{\nu'}(sc) = I_{\nu'}k(sc) \cos\theta' \, dsdAd\omega'd\nu'dt. \tag{5.4}$$

This energy is scattered into other frequencies and directions. Let the scattering be described by a phase function p. Then $p(X,\nu'\rightarrow\nu)d\nu d\omega/4\pi$ is the probability that, when a scattering does take place, the photon will be deflected through the angle X into solid angle $d\omega$ and its frequency will be changed from ν' to between ν and $\nu + d\nu$. p is normalized so that its integral over all directions and frequencies is 4π. X is the scattering angle, the angle between the incident direction along $d\omega'$ and the scattered direction along $d\omega$. If (θ,ϕ) and (θ',ϕ') are the spherical angles of the scattered and the incident directions, then

$$\cos X = \cos\theta \cos\theta' + \sin\theta \sin\theta' \cos(\phi - \phi'). \tag{5.5}$$

Notice that X is symmetric in $d\omega$ and $d\omega'$, so that the scattering angle is the same if the roles of incident and scattered photons are interchanged.

One can now find from equation (5.4) and the phase function the amount of energy scattered into the new direction and frequency from the old pencil of radiation. Noting in equation (5.4) that $\cos\theta' \, dsdA = dV$, the volume element responsible for the scattering,

$$dE_\nu(\nu'\rightarrow\nu,\omega'\rightarrow\omega) = k(sc)I_{\nu'}p(X,\nu'\rightarrow\nu) \, dVd\nu d\nu' \frac{d\omega d\omega'}{4\pi} \, dt. \tag{5.6}$$

The energy scattered into the new frequency and solid angle intervals by dV from all incident directions and frequencies is obtained by integrating equation (5.6) over all $d\omega'$ and $d\nu'$. But according to equation (3.1), this is just $j_\nu(sc) \, dVd\nu d\omega dt$. Thus the scattering part of the emission coefficient is

$$j_\nu(sc) = \iint k(sc,\nu') \, I_{\nu'}(\theta',\phi') \, p(X,\nu'\rightarrow\nu) \, \frac{d\nu'd\omega'}{4\pi} . \tag{5.7}$$

The arguments of frequency and direction have been explicitly indicated in the quantities in the integrand above. Equation (5.7) shows the direct way in which the scattering part of the emission coefficient (and therefore the source function) depends upon the radiation field.

Nothing further can be done with equation (5.7) unless the form of the phase function p is known. One cannot solve a radiation transport problem unless the physics of the interaction between matter and radiation is understood. If the scattering particles have relativistic velocities, or if the photons have energies comparable with the rest mass of the particles, then a significant frequency shift will occur and the scattering is described by the Compton effect. Although these conditions do occur in a number of astrophysically interesting situations, they do not occur in normal stellar atmospheres and they will not be considered further here. Thus the present concern is only with the case of no frequency change. (Frequency shifts due to the Doppler effect are of no importance for continuum transitions; as indicated in Chapter 4, they are very important for line transitions.) The phase function then is of the form $p(X)\delta(\nu' - \nu)$, where δ is the Dirac δ function that is zero everywhere except where the argument is zero. At the latter point the function becomes infinite in such a way that its integral is equal to unity. The frequency integral in equation (5.7) then has the effect of changing all ν' values to ν. One finds

$$j_\nu(sc) = k(sc,\nu) \int I_\nu(\theta',\phi')\ p(X)\ \frac{d\omega'}{4\pi}\ . \tag{5.8}$$

From this the source function is easily seen to be

$$S_\nu(sc) = \int I_\nu(\theta',\phi')\ p(X)\ \frac{d\omega'}{4\pi}\ . \tag{5.9}$$

Note that both the emission coefficient and the source function will generally depend upon direction (θ,ϕ).

The Rayleigh phase function, valid for ordinary Thomson scattering, is given by

$$p(X) = \frac{3}{4}\ (1 + \cos^2 X). \tag{5.10}$$

In most scattering problems, it is apparently felt that the direction dependence given by equation (5.10) is not important enough to justify the extra work necessary to take it into account. S. Chandrasekhar (*Ap. J.* **100**, p. 117, 1944) in fact showed that, at least under certain circumstances, it does not have an important effect on the radiation field. Thus one generally assumes in practice that the scattering has no direction dependence, so $p(X) = 1$, and

$$S_\nu(sc) = \int I_\nu(\theta',\phi')\ \frac{d\omega'}{4\pi} = J_\nu. \tag{5.11}$$

The scattering as represented by equation (5.11) is called isotropic.

Considering the non-scattering part of the source function, let two energy levels of an atom be designated 1 (lower) and 2 (upper). Transitions between these levels give rise to the absorption and emission of radiation and, therefore, to a contribution to the source function. There are three kinds of radiative transitions which can take place between the levels: spontaneous emission from 2 to 1; induced emission from 2 to 1; and absorption from 1 to 2. The rates at which these transitions take place are most conveniently expressed in terms of the Einstein coefficients. If B_{12}, B_{21}, and A_{21} are the Einstein coefficients of absorption, induced emission, and spontaneous emission, respectively, then

$$N(ab)\ d\nu d\omega = N_1 B_{12}\ I_\nu\ d\nu\ \frac{d\omega}{4\pi}\ ,$$

$$N(ie)\ d\nu d\omega = N_2 B_{21}\ I_\nu\ d\nu\ \frac{d\omega}{4\pi}\ , \qquad (5.12)$$

$$N(se)\ d\nu d\omega = N_2 A_{21}\ d\nu\ \frac{d\omega}{4\pi}\ .$$

The left sides of the above equations represent the number of transitions (ab = absorption, ie = induced emission, se = spontaneous emission) per unit volume per unit time in which the photon which is absorbed or emitted comes from or goes into frequency interval $d\nu$ and solid angle $d\omega$. N_1 and N_2 are the numbers of atoms per unit volume in these energy levels. If E_{12} is the energy difference between the levels, then the frequency ν appearing in equation (5.12) must be the atomic constant E_{12}/h, where h is Planck's constant, and is not an independent physical parameter. The Einstein coefficients as given by equations (5.12) are atomic constants that depend on the atom and the levels in question, but they do not depend upon temperature or other physical conditions. The coefficients are sometimes defined in a different way, such as in terms of energy density rather than from intensity, and the effect is simply to change them by some multiplying constant.

If the levels belong to the continuum, then the abundances N_1 and N_2 are the number of electron-ion pairs per unit relative kinetic energy per unit volume, so they are proportional to the density of free electrons as well as the ion density. If both levels are bound so that the transition is a line, then a suitable broadening function, as discussed in Chapter 4, must also be included if one wishes to keep the Einstein coefficients strictly independent of physical conditions.

Consider the volume element $dV = \cos\theta\ dA ds$, where as usual ds makes the angle θ with the normal to dA. The matter within dV modifies the energy in the pencil of radiation propagating along ds and within solid angle $d\omega$. Since energy gained or lost equals number of transitions times $h\nu$, the net energy gained by the pencil in time dt due to the volume dV is found from equation (5.12) to be

$$dE_\nu = \frac{h\nu}{4\pi}(N_2A_{21} + N_2B_{21}I_\nu - N_1B_{12}I_\nu)\, dVd\nu d\omega dt. \tag{5.13}$$

The energy change in the pencil of radiation results in a differential change of intensity. This change is found from equation (1.1) to be

$$dE_\nu = dI_\nu \cos\theta\ dAd\omega d\nu dt\ . \tag{5.14}$$

If expressions (5.13) and (5.14) are equated, one obtains the following form of the equation of transfer:

$$\frac{dI_\nu}{ds} = \frac{h\nu}{4\pi} N_2A_{21} - \frac{h\nu}{4\pi}(N_1B_{12} - N_2B_{21})I_\nu. \tag{5.15}$$

By comparing this with equation (4.1), one finds the following expressions for the absorption coefficient and the source function:

$$k(ns) = (N_1B_{12} - N_2B_{21})\frac{h\nu}{4\pi}\ , \tag{5.16}$$

$$S_\nu(ns) = \frac{N_2A_{21}}{N_1B_{12} - N_2B_{21}}. \tag{5.17}$$

The non-scattering source function due to this one transition depends only on the ratio of the populations N_1/N_2 in addition to the values of the Einstein coefficients. Thus, the non-scattering part of the source function depends on whatever fixes the above population ratio.

Assume that conditions of thermodynamic equilibrium (TE) hold. In TE the radiation field is uniform, so $dI_\nu/ds = 0$, and the intensity equals the source function. But in TE the intensity is given by the Planck function:

$$B_\nu(T) = \frac{2h\nu^3/c^2}{e^{h\nu/kT} - 1}\ . \tag{5.18}$$

Also, in TE the population ratio is given by the Boltzmann relation:

$$\frac{N_1}{N_2} = \frac{g_1}{g_2}e^{h\nu/kT} = \frac{g_1}{g_2}e^{E_{12}/kT}\ . \tag{5.19}$$

Equations (5.17)-(5.19) then lead to the following:

$$\frac{A_{21}}{B_{21}} = \frac{2h\nu^3}{c^2}\ , \qquad \frac{B_{12}}{B_{21}} = \frac{g_2}{g_1}\ . \tag{5.20}$$

The **g**'s are the statistical weights of the levels, given by the number of separate quantum mechanical states having the given energy.

Although equations (5.20) were derived by assuming that TE holds, the Einstein coefficients are atomic constants independent of that assumption. Thus relations (5.20) are universally valid. The source function in equation (5.17) can be expressed, using equation (5.19), as follows:

$$S_\nu(ns) = \frac{2h\nu^3/c^2}{(g_2N_1/g_1N_2) - 1}. \tag{5.21}$$

Although equation (5.19) is valid only in TE, one can define what is known as an excitation temperature T_{ex} by the realtion

$$\frac{N_1}{N_2} = \frac{g_1}{g_2} e^{h\nu/kT_{ex}} . \tag{5.22}$$

It follows that

$$S_\nu(ns) = B_\nu(T_{ex}) . \tag{5.23}$$

Thus the non-scattering part of the source function is equal to the Planck function of the excitation temperature.

If conditions are very far from TE, the excitation temperature will have little semblance to other measures of temperature, and equation (5.23) has little physical content. If conditions are not too far from TE, however, T_{ex} will not differ much from other significant measures of temperature and equation (5.23) becomes quite meaningful in terms of these measures.

The kinetic temperature T_k describing the velocity distribution of the free particles is an important measure of temperature. Under most astrophysical conditions elastic collisions are sufficiently numerous that T_k is a well defined quantity. If conditions are not too far from TE, T_{ex} for any given set of levels will be close to T_k, so the non-scattering part of the source function becomes $B_\nu(T_k)$. Whenever this is a sufficiently good approximation for all transitions of importance, the condition of local thermodynamic equilibrium (LTE) is said to hold.

In LTE, the source function is independent of the particular atomic transitions that are involved. It depends only on the kinetic temperature. Thus, as far as the source function is concerned, LTE is the same as TE. The radiation field in LTE, however, may be quite different from that in pure TE.

The analysis of a given situation to determine the validity of LTE is very complicated. The distribution of particles over the different energy levels is determined by both collisional and radiative transitions. The collisional effects are determined by the kinetic properties of the matter, so collisions tend to produce LTE. The radiative transitions are much more complicated. They would also drive the system toward LTE if the photons were incident uniformly from all directions and if the photons were in equilibrium with the local kinetic temperature. These two conditions are met in the deep interior of a star, but the atmosphere of a star is a place where, by definition, these two conditions are not met. Thus, it becomes necessary to solve the detailed equations of statistical equilibrium coupled with the equation of transfer in order to test any given case.

The consensus of opinion is that for a majority of stars LTE for continuum processes is a very good approximation. The main exceptions are extremely hot and extremely cool stars, very high luminosity stars, and stars with extended atmospheres. A good review paper by B.E.J. Pagel, *Proc. Roy. Soc.* A. **306**, 91, 1968, is recommended for further reading on this question. The line problem is even more complicated and will be considered separately in Chapter 4.

Unless otherwise stated, it will be assumed that the non-scattering part of the continuum source function is in LTE and that the scattering part is isotropic. Thus the total continuum source function is

$$S_\nu = qB_\nu(T) + (1 - q)J_\nu, \tag{5.24}$$

where $q = k(ns)/k$ is the ratio of non-scattering to total absorption. T without subscript is defined to mean the kinetic temperature.

In TE the intensity is equal to the Planck function. In this case, both scattering and non-scattering parts of the source function are given by the Planck function. This condition will be closely approached as one goes very deeply into the star.

For most stars $k(ns)$ is far greater than $k(sc)$, q is thus very close to unity and the total continuum source function is that of LTE. In very hot stars, however, the large number of free electrons coming from the ionization of hydrogen causes electron scattering to bring $k(sc)$ up to a significant value, and both terms in equation (5.24) must be included.

6. Special Integrals for Plane Media

Most stellar atmospheres are extremely thin compared with the radii of the stars. Except when considering the extreme edge of the disk, therefore, one can ignore the curvature of the atmosphere when studying its radiation transport. This is a very great simplification in the geometry, as the equation of transfer can then be expressed in terms of the derivative of only one coordinate.

It is standard to measure distances positive into the atmosphere of a star, but to measure angles from the outward direction. This convention introduces a negative sign into the equation of transfer. Let x be the linear distance into the star measured from the surface, and let θ be the angle a direction makes with the outward direction. Then if ds is directed along θ, one has

$$dx = -\cos\theta \; ds. \tag{6.1}$$

The equation of transfer (4.1) then becomes

$$\cos\theta \, \frac{dI_\nu}{dx} = kI_\nu - j_\nu = k(I_\nu - S_\nu). \tag{6.2}$$

Replace x as the independent variable by the optical depth τ defined as

$$d\tau = k dx. \tag{6.3}$$

τ is also taken as zero at the surface of the star, and it is a measure of the perpendicular distance below the surface. The transfer equation (6.2) then becomes

$$\mu \frac{dI_\nu}{d\tau} = I_\nu - S_\nu. \tag{6.4}$$

where $\mu = \cos\theta$.

The solution of equation (6.4) is indicated by obvious modifications of equation (4.7). The result is most easily expressed if a distinction is made between upward and downward directions. This is caused by the different boundary conditions in the two directions, the atmosphere extending to infinity downward. The quantity $\mu = \cos\theta$, unless otherwise specified, will remain positive; the downward direction will be indicated by a negative sign in front. Thus $I_\nu(+\mu)$ and $I_\nu(-\mu)$ are the intensities directed into the separate hemispheres. One then obtains

$$I_\nu(\tau,+\mu) = \int_\tau^\infty S_\nu(\tau') \, e^{-(\tau'-\tau)/\mu} \frac{d\tau'}{\mu},$$

$$(0 \leq \mu \leq 1) \tag{6.5}$$

$$I_\nu(\tau,-\mu) = I_\nu(0,-\mu) \, e^{-\tau/\mu} + \int_0^\tau S_\nu(\tau') \, e^{-(\tau-\tau')/\mu} \frac{d\tau'}{\mu}.$$

$I_\nu(0,-\mu)$ is the intensity incident on the top of the atmosphere. In most problems this is zero. In agreement with equation (5.24), the source function has been assumed to be a function of depth but not of direction.

The mean intensity, flux, and radiation pressure can also be found in terms of integrals of the source function over the atmosphere. If the relations (6.5) are multiplied by the appropriate expressions, the integrals of equations (1.3), (1.6), and (1.10) can be carried out in a straightforward fashion. The results are given in terms of the exponential-integral function $E_n(y)$:

$$E_n(y) = \int_1^\infty e^{-yu} u^{-n} \, du = \int_0^1 e^{-y/v} v^{n-2} \, dv, \tag{6.6}$$

$$J_\nu(\tau) = \frac{1}{2}\int_0^\infty S_\nu(\tau') E_1(|\tau' - \tau|) \, d\tau', \tag{6.7}$$

$$F_\nu(\tau) = 2\int_\tau^\infty S_\nu(\tau') E_2(\tau' - \tau) \, d\tau' - 2\int_0^\tau S_\nu(\tau') E_2(\tau - \tau') d\tau', \tag{6.8}$$

$$P_{r_\nu}(\tau) = \frac{2\pi}{c}\int_0^\infty S_\nu(\tau')E_3(|\tau' - \tau|)\, d\tau' . \tag{6.9}$$

The exponential-integral functions are somewhat similar to the ordinary exponential function. $E_n(0) = 1/(n - 1)$, and $E_n(y) \rightarrow e^{-y}/y$ as $y \rightarrow \infty$. It is seen that E_1, E_2, and E_3 take the part of the "exponential attenuation" for J_ν, F_ν, and P_{r_ν} respectively. In equations (6.7)-(6.9) the radiation incident on the top of the atmosphere has been ignored.

An important energy conservation relation can be obtained from the flux by taking the divergence of equation (1.4). Since the unit vector $\mathbf{s}$ and $d\omega$ do not depend upon position, only the intensity is operated upon:

$$\nabla\cdot\mathbf{F}_\nu = \int \mathbf{s}\cdot\nabla I_\nu d\omega. \tag{6.10}$$

But the integrand is simply the component of the gradient of the intensity along $\mathbf{s}$, which is dI_ν/ds. From the equation of transfer this is seen to be $k(S_\nu - I_\nu)$. The integral over the solid angle then yields

$$\nabla\cdot\mathbf{F}_\nu = 4K(S_\nu - J_\nu). \tag{6.11}$$

In a plane medium the vector flux is directed upward out of the star, so the divergence becomes $-d/dx$. Using equation (6.3), one obtains

$$\frac{dF_\nu}{d\tau} = 4(J_\nu - S_\nu). \tag{6.12}$$

Equations (6.11) and (6.12) state that the flux at any frequency results from the excess of emissions over absorptions at that frequency.

If the transfer equation (6.4) is multiplied through by μ and the result integrated over the solid angle, one finds

$$\frac{dP_{r_\nu}}{d\tau} = \frac{\pi}{c} F_\nu. \tag{6.13}$$

Again the assumption is used that the source function does not depend upon direction. If another derivative of equation (6.13) with respect to optical depth is taken, and if equation (6.12) is substituted into the result, then

$$\frac{d^2P_{r_\nu}}{d\tau^2} = \frac{4\pi}{c}(J_\nu - S_\nu). \tag{6.14}$$

These relations will be referred to in later sections.

2

THE GRAY ATMOSPHERE

7. Introduction

In the present chapter the concern will be with atmospheres which have the following characteristics: 1. static; 2. semi-infinite; 3. plane; 4. no radiation incident on top; 5. LTE; 6. energy carried only by radiation; 7. absorption coefficient independent of frequency. A gray atmosphere is usually defined as one which satisfied condition 7 alone, although in this chapter the term gray is used, for convenience, to indicate all seven conditions, unless otherwise stated.

The first five conditions above need no elaboration at present beyond the discussion in Chapter 1. Condition 6 is generally called radiative equilibrium. It means that radiative energy is conserved, so that whatever radiative energy is put into the bottom of the atmosphere must also come out the top. For a plane atmosphere, the radiative flux is a constant independent of depth. Radiative equilibrium is a good approximation for most atmospheres, as the nuclear energy sources are far removed from the atmosphere, and radiation is usually the dominant form of energy transport. There are some cases, however, when convection may carry a significant fraction of the energy through part of the atmosphere. Then the exchange of energy between radiation and convection must be taken into account. The problem of convection will be discussed in the next chapter.

The assumption that the absorption coefficient is independent of frequency is highly artificial. Historically this assumption was often made because it represented a tremendous simplification and because it offered a large amount of physical insight on the more complicated problems of real atmospheres. Besides, the correct variation of absorption with frequency was not well known. It is now known that the absorption throughout most of the important wavelength regions of moderate temperature stars is dominated by the negative ion of hydrogen, and H^- absorption is very nearly independent of frequency. Thus the gray atmosphere turns out to be a much better representation of real stars than one could have expected.

With LTE the equation of transfer (6.4) becomes

$$\mu \frac{dI_\nu}{d\tau} = I_\nu - B_\nu, \tag{7.1}$$

where τ is now independent of frequency. This relation can now be integrated over all frequencies (which obviously cannot be done in the non-gray case) to yield a relation between integrated quantities:

$$\mu \frac{dI}{d\tau} = I - B. \tag{7.2}$$

The integrated Planck function is given by

$$B = \frac{\sigma}{\pi} T^4, \tag{7.3}$$

where σ is the Stefan-Boltzmann radiation constant (5.67×10^{-5} erg cm^{-2} sec^{-1} $^oK^{-4}$).

In the gray case the flux integral (6.12) can also be integrated over frequency. For a plane atmosphere in radiative equilibrium the flux is independent of depth. Thus $dF/d\tau$ is zero, and it follows that

$$B = J. \tag{7.4}$$

Note that J_ν does not equal B_ν. Equations (6.7) and (6.8) integrated over frequency then give the following for mean intensity and flux:

$$J(\tau) = B(\tau) = \frac{1}{2}\int_0^\infty B(\tau')E_1(|\tau' - \tau|)\, d\tau'; \tag{7.5}$$

$$F(\tau) = \text{constant} = 2\int_\tau^\infty B(\tau')E_2(\tau' - \tau)\, d\tau' - 2\int_0^\tau B(\tau')E_2(\tau - \tau')\, d\tau'. \tag{7.6}$$

The effective temperature T_e of a star is the temperature of a black body which radiates energy at the same rate per unit area as the star. In terms of the surface flux one has

$$F = \frac{\sigma}{\pi} T_e^4 . \tag{7.7}$$

Equations (7.5) and (7.6) suffice to uniquely determine $B(\tau)$ for a given value of F (or T_e). But B is a known function of temperature through equation (7.3), so the temperature-optical depth relation is also determined. Then the monochromatic source function B_ν is a known function of optical depth, and equations (6.5) fix the monochromatic intensity for all depths and directions. In terms of the single input parameter, F or T_e, the entire radiation field of a gray atmosphere is determinable.

It may seem surprising that values of the absorption coefficient k are not needed in order to find the radiation field. This is essentially due to two factors: 1) optical depth rather than linear depth is used as the independent variable; and 2) the absorption is known to be the same at all frequencies. If one wants the solution in terms of the linear depth x instead of optical depth, then k must be known and the problem becomes much more complicated.

If k is large, most of the energy escaping from a star will come from regions which are very close in linear distance to the surface; if k is small, regions at a much greater linear distance from the surface will be important. In both cases the region near optical depth $\tau = 1$ makes the most contribution to the emitted radiation. If k changes with frequency, a region may be at optical depth unity at one frequency and at a much greater or smaller optical depth at another frequency. In order to determine the energy balance over all frequencies, it is necessary to tie together these different regions through the absorption coefficient, i.e., k must be known. But when k is known to be independent of frequency, optical depth unity at one frequency is the same for any other frequency, and further information about k is not needed.

The exact solution for the gray temperature distribution was given by C. Mark, *Phys. Rev.* **72**, 558, 1947. In the next section a very useful approximation method will be applied to the gray problem, and this will be followed by a more accurate method of approach. In both cases the methods have many applications beyond the simple gray problem considered here.

8. The Eddington Approximation

Consider the quantity defined as follows:

$$Q_n(\tau) = \frac{1}{2}\int_0^\infty S_\nu(\tau')E_n(|\tau' - \tau|)\,d\tau', \tag{8.1}$$

where E_n is the exponential-integral function given by equation (6.6). If one makes the substitution $\tau' = \tau + y$, equation (8.1) becomes

$$Q_n(\tau) = \frac{1}{2}\int_0^\tau \Big[S_\nu(\tau + y) + S_\nu(\tau - y)\Big]E_n(y)\,dy + \frac{1}{2}\int_\tau^\infty S_\nu(\tau + y)\,E_n(y)\,dy. \tag{8.2}$$

Since E_n decreases rapidly with increasing values of the argument, most of the contribution to Q comes from small values of y. For very large values of τ, therefore, the second term in equation (8.2) is negligible compared to the first and the integrand in the first term can be expanded in a first order Taylor series: $S_\nu(\tau \pm y) = S_\nu(\tau) \pm y\,dS_\nu/d\tau$. Thus for very large τ

$$Q_n(\tau) = S_\nu(\tau)\int_0^\tau E_n(y)\,dy = \frac{1}{n}S_\nu(\tau). \tag{8.3}$$

The second part of equation (8.3) follows from the properties of $E_n(y)$. It is easily seen from equation (8.1) that $J_\nu = Q_1$ and $P_{r_\nu} = (4\pi/c)Q_3$. Equation (8.3) then indicates that, in the limit of very large optical depths,

$$P_{r_\nu}(\tau) = \frac{4\pi}{3c}J_\nu(\tau). \tag{8.4}$$

The differential equation (6.14) then gives

$$\frac{d^2J_\nu}{d\tau^2} = 3(J_\nu - S_\nu). \tag{8.5}$$

Again, equations (8.4) and (8.5) are valid in the limit of very large optical depths. The Eddington approximation consists of the assumption that these relations are valid for all depths, including the surface. It should be noted that nothing has been said about radiative equilibrium, LTE, or the frequency dependence of the absorption coefficient. The Eddington approximation is quite independent of the assumption of grayness.

Approximate boundary conditions may be found in the following way: The surface flux is $F_\nu(0) = 4Q_2(0)$. Equation (8.3) suggests that this is roughly equal to $2Q_1(0) = 2J_\nu(0)$, so

$$J_\nu(0) = \frac{1}{2}F_\nu(0), \tag{8.6}$$

$$\left.\frac{dJ_\nu}{d\tau}\right)_0 = \frac{3}{2}J_\nu(0). \tag{8.7}$$

Sometimes the factor $\sqrt{3}$ is used in equation (8.7) instead of 3/2 (M.Krook, *Ap. J.* **122**, 488, 1955), but the distinction is insignificant on the present level of approximation. Equation (8.7) is derived by differentiating equation (8.4) with respect to the optical depth and equating the result to the radiation pressure relation (6.13).

The Eddington approximation will now be applied to the gray atmosphere. As optical depth is independent of frequency, equation (8.5) can be integrated over frequency. When combined with equation (7.4) it gives

$$\frac{d^2J}{d\tau^2} = 0. \tag{8.8}$$

The solution is

$$J = B = a\tau + b\ , \tag{8.9}$$

where a and b are constants. Condition (8.7) requires $a = (3/2)b$, while equation (8.6) gives $b = F/2$. The depth dependence of J and B is then

$$J = B = \frac{3}{4} F(\tau + \frac{2}{3}). \tag{8.10}$$

Equation (8.10) can be written in terms of the temperature by using equations (7.3) and (7.7),

$$T^4 = \frac{3}{4} T_e^4 (\tau + \frac{2}{3}). \tag{8.11}$$

The surface temperature T_0 is seen to be $(1/2)^{1/4} = 0.841$ times the effective temperature.

Equation (8.11) gives the temperature at each depth, so equations (6.5) can then be used to find the monochromatic intensity for any depth and direction. As indicated earlier, the entire radiation problem has been solved (in a certain approximation) in terms of the single parameter F (or T_e). Equation (8.11) shows how the temperature distribution scales for atmospheres of different effective temperatures.

Of particular interest is the surface intensity in integrated radiation:

$$I(0,+\mu) = \frac{3}{4} F(\mu + \frac{2}{3}). \tag{8.12}$$

By comparing equations (8.10) and (8.12), it is seen that the surface intensity is equal to the source function evaluated at an optical depth equal to $\cos\theta$, i.e., at an optical depth measured along the line of sight of unity. This is a very useful relation when only rough accuracy is sufficient. Finally, the quantity $\phi(\mu) \equiv I(0,+\mu)/I(0,1)$ is known as the limb darkening function:

$$\phi(\mu) = \frac{3}{5} \mu + \frac{2}{5} . \tag{8.13}$$

One can check the accuracy of the Eddington approximation for the gray problem by substituting the solution (8.10) into equations (7.5) and (7.6). The exact solution would give the identities $B(\tau) = B(\tau)$ and $F = F$. The solution (8.10) yields:

$$B(\tau) = B(\tau) \left[1 + \frac{3E_3(\tau) - 2E_2(\tau)}{3\tau + 2} \right],$$

$$F = F \left[1 + E_3(\tau) - \frac{3}{2} E_4(\tau) \right]. \tag{8.14}$$

The terms containing the exponential integrals represent deviations from the exact solution arising from the approximation. The errors as functions of optical depth are listed below:

τ	$\dfrac{3E_3(\tau) - 2E_2(\tau)}{3\tau + 2}$	$E_3(\tau) - \frac{3}{2}E_4(\tau)$
0.0	-0.250	0.000
0.1	-0.085	-0.015
0.2	-0.036	-0.022
0.5	+0.003	-0.026
1.0	0.006	-0.019
2.0	0.002	-0.007

As one would expect, the errors are appreciable only near the surface. They rapidly go to zero as the optical depth becomes significantly greater than unity.

9. The Method of Discrete Ordinates

The discussion in this section is based largely on the development by S. Chandrasekhar in his book *Radiative Transfer*. Equations (7.2) and (7.4) indicate that the equation of transfer for the gray atmosphere can be written

$$\mu\frac{dI}{d\tau} = I - \frac{1}{2}\int_{-1}^{+1} I(\tau,\mu')\,d\mu'. \tag{9.1}$$

The integration over the azimuth angle ϕ has already been carried out, as in the plane atmosphere, I does not depend upon ϕ. The boundary conditions are that there is no radiation incident on the surface: $I(0,-\mu) = 0$; and that no radiation can pass through the infinite optical thickness without being absorbed: $I(\tau)e^{-\tau} \rightarrow 0$ as $\tau \rightarrow \infty$.

Chandrasekhar's method of solving the integral-differential equation (9.1) is to replace the integral by a summation. The technique is chosen so that the maximum accuracy is obtained for a given number of terms in the summation. The Gaussian quadrature method suggests itself,and a brief digression on this method at this stage will be useful.

The m-point quadrature formula for use with equation (9.1) is

$$\int_{-1}^{+1} f(\mu)d\mu \simeq \sum_{i=1}^{m} a_i f(\mu_i). \tag{9.2}$$

The Gaussian method consists of taking for the points of division μ_i the zeros of the mth order Legendre polynomial:

$$P_m(\mu_i) = 0, \quad i = 1,2,\ldots,m. \tag{9.3}$$

The weights a_i are given by

$$a_i = \frac{1}{P'_m(\mu_i)}\int_{-1}^{+1} \frac{P_m(\mu)}{\mu - \mu_i}\,d\mu. \tag{9.4}$$

P'_m is the derivative of P_m. For these choices of a_i and μ_i, the integral is given exactly by the sum in equation (9.2) if $f(\mu)$ is a polynomial of degree $(2m-1)$ or less. For more general functions the Gaussian sum gives a far better representation of the integral, for a given number of terms, than the usual techniques which have the μ_i equally spaced. It is the most accurate method for given m for functions which are well represented by polynomials.

It is convenient to restrict m to even integers, $m = 2n$. In this case the Legendre polynomial $P_{2n}(\mu)$ contains only even powers of μ, so the roots occur in pairs $\pm\mu_i$, and the directions represented by the μ_i are symmetric in the outer and inward hemispheres. From equation (9.4) one can show that the weight associated with $+\mu_i$ is the same as that with $-\mu_i$. Thus it becomes convenient to designate the weights and points by (a_i, μ_i),

$$a_i = a_{-i}, \qquad \mu_i = -\mu_{-i} \quad i = \pm 1, \pm 2, \ldots, \pm n. \tag{9.5}$$

The first three sets of weights and points are given below:

$n = 1$	$\mu_{\pm 1} = \pm 0.577350$	$a_{\pm 1} = 1.000000$
$n = 2$	$\mu_{\pm 1} = \pm 0.339981$	$a_{\pm 1} = 0.652145$
	$\mu_{\pm 2} = \pm 0.861136$	$a_{\pm 2} = 0.347855$
$n = 3$	$\mu_{\pm 1} = \pm 0.238619$	$a_{\pm 1} = 0.467914$
	$\mu_{\pm 2} = \pm 0.661209$	$a_{\pm 2} = 0.360762$
	$\mu_{\pm 3} = \pm 0.932470$	$a_{\pm 3} = 0.171324$

Since equation (9.2) is exact if $f(\mu)$ is a polynomial of degree $(4n - 1)$ or less, an important property of the Gaussian weights and points is

$$\int_{-1}^{+1} \mu^p \, d\mu = \sum_{i=\pm 1}^{\pm n} a_i \mu_i^{\,p} = \frac{2}{p+1} \quad (p \text{ even})$$

$$p \leq (4n - 1) \tag{9.6}$$

$$= 0. \qquad (p \text{ odd})$$

The application of this method to the gray problem begins with the replacement of the integral in the equation of transfer (9.1) by a Gaussian sum of order n :

$$\mu \frac{dI}{d\tau} = I - \frac{1}{2} \sum_{j=\pm 1}^{\pm n} a_j I_j \, , \tag{9.7}$$

where I_j is the notation for $I(\tau, \mu_j)$. To the extent that the above sum is a good representation of the integral in equation (9.1), equation (9.7) is valid for all depths and directions.

Equation (9.7) is linear, homogeneous differential equation, so a solution of the following form will be sought:

$$I(\tau,\mu) = g(\mu)e^{-k\tau}, \tag{9.8}$$

where k is a constant. If this is substituted into equation (9.7), the condition on $g(\mu)$ is

$$g(\mu) = \frac{\frac{1}{2}\sum_{i=\pm 1}^{\pm n} a_i g_i}{1 + k\mu} = \frac{\text{constant}}{1 + k\mu}. \tag{9.9}$$

If equation (9.9) is evaluated at $\mu = \mu_j$, each term multiplied by a_j and summed over all j values, the following condition on k is obtained:

$$\sum_{j=\pm 1}^{\pm n} \frac{a_j}{1 + k\mu_j} = 2. \tag{9.10}$$

By using conditions (9.5), one can transform the above to

$$\sum_{j=1}^{n} \frac{a_j}{1 - k^2\mu_j^2} = 1. \tag{9.11}$$

The sum in equation (9.11) is taken only over positive values of j.

It is apparent that equation (9.11) is equivalent to a polynomial equation of degree n in k^2, and there are thus 2n roots for k which occur in ± pairs. One of the solutions is $k^2 = 0$, however, as can be seen from equation (9.6) for $p = 0$. Thus there are only $(2n - 1)$ separate solutions of the form (9.8). The $(2n - 2)$ non-zero values of k which satisfy equation (9.11) will be designated as k_m, $m = \pm 1, \ldots, \pm(n-1)$, with $k_m = -k_{-m}$. The $k = 0$ solution will be denoted as Q. The 2nth independent solution of equation (9.7) is easily seen by inspection to be the function $(\tau + \mu)$. The general solution of equation (9.7) is then a linear sum of the 2n special solutions:

$$I(\tau,\mu) = \text{constant}\left[\sum_{m=\pm 1}^{\pm(n-1)} \frac{L_m e^{-k_m\tau}}{1 + k_m\mu} + Q + \tau + \mu\right]. \tag{9.12}$$

The L_m, Q, and the multiplying constant in equation (9.12) are the 2n constants which must be evaluated from the physics of the problem and the boundary conditions. The boundary condition at large depths is that $I(\tau)e^{-\tau}$ go to zero at sufficiently large optical depths. This condition is violated in equation (9.12) by the terms in $e^{-k_m\tau}$ for negative values of m, as one can show that $k_m^2 > 1$ for all m. Thus the lower boundary condition requires that

$$L_m = 0, \quad m = -1, -2, \ldots, -(n-1). \tag{9.13}$$

The solution for the intensity can now be written as:

$$I(\tau,\mu) = \text{constant}\left[\sum_{m=1}^{n-1} \frac{L_m e^{-k_m\tau}}{1 + k_m\mu} + Q + \tau + \mu\right]. \tag{9.14}$$

The upper boundary condition is that the radiation incident on the top of the atmosphere be zero:

$$\sum_{m=1}^{n-1} \frac{L_m}{1 - k_m\mu} + Q - \mu = 0, \qquad 0 < \mu \leq 1. \tag{9.15}$$

It is immediately seen that equation (9.15) cannot be satisfied, for the n constants L_m and Q are not sufficient to insure that the equation is satisfied for all μ. The best that one can do is to fix n values of μ and satisfy the upper boundary condition only for these n directions. Thus (9.14) represents a self-consistent solution for all upward directions, but this is true only for n discrete downward directions. One obviously should choose the $-\mu_i$, $i = 1,\ldots,n$ as the n special downward directions for which the boundary conditions are to be satisfied, as this is the most accurate representation for the sum appearing in equation (9.7). Thus instead of equation (9.15), the L_m and Q are determined by requiring the following:

$$\sum_{m=1}^{n-1} \frac{L_m}{1 - k_m\mu_i} + Q - \mu_i = 0, \qquad i = 1,\ldots,n. \tag{9.16}$$

The multiplying constant in equation (9.14) is related to the flux. After an integration over ϕ, one has

$$F(\tau) = 2\int_{-1}^{+1} I(\tau,\mu)\mu d\mu \simeq 2\sum_{i=\pm1}^{\pm n} a_i\mu_i I(\tau,\mu_i). \tag{9.17}$$

When equation (9.14) is substituted into this relation and the condition (9.6) is used, the result is

$$F(\tau) = 2\ \text{constant}\left[\sum_{m=1}^{n-1} L_m e^{-k_m\tau}\left(\sum_{i=\pm1}^{\pm n} \frac{a_i\mu_i}{1 + k_m\mu_i}\right) + \frac{2}{3}\right]. \tag{9.18}$$

But from equations (9.6) and (9.10),

$$\sum_{i=\pm1}^{\pm n} \frac{a_i\mu_i}{1 + k_m\mu_i} = \frac{1}{k_m}\sum_{i=\pm1}^{\pm n} a_i\left(1 - \frac{1}{1 + k_m\mu_i}\right) = 0. \tag{9.19}$$

Thus the multiplying constant is found to be $3F/4$. The intensity is then given by

$$I(\tau,\mu) = \frac{3}{4} F\left[\sum_{m=1}^{n-1} \frac{L_m e^{-k_m\tau}}{1 + k_m\mu} + Q + \tau + \mu\right]. \qquad (9.20)$$

It should again be emphasized that equation (9.20) can be applied to all upward directions, but only those downward directions along the points of division satisfy the boundary conditions. In fact equation (9.20) is seen to be singular for those directions given by $\mu = -1/k_m$, but these are not the directions of the $-\mu_i$.

The temperature distribution in the atmosphere can be found from the evaluation of the mean intensity.

$$J(\tau) = B(\tau) = \frac{1}{2}\int_{-1}^{+1} I(\tau,\mu)\, d\mu \simeq \frac{1}{2} \sum_{i=\pm 1}^{\pm n} a_i I(\tau,\mu_i). \qquad (9.21)$$

If this is evaluated from equation (9.20), the result is

$$J(\tau) = B(\tau) = \frac{3}{4} F\left[\sum_{m=1}^{n-1} L_m e^{-k_m\tau} + \tau + Q\right]. \qquad (9.22)$$

If this result is compared with equation (8.10), it is seen that the Eddington approximation corresponds to $L_m = 0$ and $Q = 2/3$. In the first three approximations, the values of the various constants are as given below:

$n = 1$	$Q = 0.577350$		
$n = 2$	$Q = 0.694025$	$k_1 = 1.972027$	$L_1 = -0.116675$
$n = 3$	$Q = 0.703899$	$k_1 = 1.225211$ $k_2 = 3.202945$	$L_1 = -0.025304$ $L_2 = -0.101245$

Equation (9.22) is commonly written in the form

$$B(\tau) = \frac{3}{4} F\left[\tau + q(\tau)\right]. \qquad (9.23)$$

According to V. Kourganoff in the book *Basic Methods in Transfer Problems*, the exact values of $q(\tau)$ are the following:

τ	$q(\tau)$	τ	$q(\tau)$
0.00	0.57735	0.5	0.68029
.05	.61076	1.0	.69854
.10	.62792	1.5	.70513
.20	.64955	2.0	.70792
.30	.66336	∞	.71045

The fact that $q(\tau)$ is very nearly constant illustrates the accuracy of the eddington approximation.

If one wishes to have an expression for the intensity valid for all directions, he can use equation (9.22) as the source function and evaluate the integrals in equation (6.5). The result agrees exactly with equation (9.20) for all outward directions; however, for negative μ a different expression is obtained:

$$I(\tau,-\mu) = \frac{3}{4} F \sum_{m=1}^{n-1} \frac{L_m}{1 - k_m\mu}(e^{-k_m\tau} - e^{-\tau/\mu}) + \tau + (Q - \mu)(1 - e^{-\tau/\mu}),$$

$$0 < \mu \leq 1. \qquad (9.24)$$

The singularities at $\mu = -1/k_m$ have been removed from this expression; note also that equation (9.24) satisfies the boundary condition $I(0,-\mu)=0$ for all inward directions.

Although the quantities considered so far are integrated over frequency, monochromatic quantities can also be found. Equation (9.22) fixes the temperature distribution, and the application of equations (6.5) then determines the intensity $I_\nu(\tau,\mu)$, as the monochromatic source fumction is now known at all depths. The integrated flux F is constant with depth in the present problem, so the increase of temperature into the star means that the monochromatic flux $F_\nu(\tau)$ has its maximum shifted toward shorter wavelengths as one considers larger optical depths.

10. Isotropic Scattering

For isotropic scattering without non-scattering, the source function is

$$S_\nu = J_\nu. \qquad (10.1)$$

Thus the equation of transfer in a plane atmosphere is

$$\mu \frac{dI_\nu}{d\tau} = I_\nu - J_\nu. \qquad (10.2)$$

This is identical in form with equations (7.2) and (7.4) for the gray problem, the only difference being that equation (10.2) is in terms of monochromatic quantities. The boundary conditions are also the same, so $I(\tau)$, $J(\tau)$, etc., which are solutions of the gray atmosphere must be identical to $I_\nu(\tau)$, $J_\nu(\tau)$, etc., the solutions of the isotropic scattering problem. Thus the relations found in the last section can also be applied here, and the isotropic scattering atmosphere has already been solved in the plane, semi-infinite case. The fact that optical depth may depend upon frequency in the scattering case does not change the mathematical form of the solution.

In the gray problem, the source function was equal to the integrated mean intensity J only for radiative equilibrium, i.e., only if F = constant. Thus equation (10.2) must also require F_ν = constant. A term by term integration of equation (10.2) over solid angle shows that this is indeed true. It is apparent on physical grounds that the monochromatic flux must be constant for the scattering problem in a plane atmosphere. The fact that the scattering process does not change the frequency of a photon and that scattering is the only thing that can happen to a photon mean that there is no coupling between different frequencies or with other forms of energy. Whatever radiative flux is put into the bottom of the atmosphere at any frequency must come out the top at the same frequency. Each frequency is transported independently of the others.

The gray atmosphere in radiative equilibrium and the isotropic scattering problem are seen to be quite different physically, but they have the same mathematical solution. In the gray problem, energy absorbed at one frequency can be emitted at another, and the temperature is distributed with optical depth in such a way that the total flux carried in all frequencies is a constant. There is only one independent parameter, the integrated flux (or the effective temperature), and the solution easily scales from one flux to another.

In the scattering problem, temperature is irrelevant. The energy absorbed at one frequency is emitted at the same one, and the source function must find a distribution with depth such that the flux at the given frequency is constant. There is one independent parameter for each frequency, namely F_ν.

There are more general types of scattering problems which have been attacked by sophisticated mathematical methods, but these are not considered here. One can consult the following for further work in this area: I.W. Busbridge, *The Mathematics of Radiative Transfer*, Cambridge, 1961; V.V. Sobolev, *A Treatise on Radiative Transfer*, Van Nostrand, 1963; and Chandrasekhar's *Radiative Transfer*.

3

THE NON-GRAY ATMOSPHERE

11. The Model Atmosphere

When the absorption coefficient depends on frequency, the procedures of the last chapter are not successful in determining the radiation field in general. As is pointed out in Section 7, it is necessary to know the relative absorptions at different frequencies, and this usually means one must find values of κ itself. (It is usually more convenient to use the mass rather than the volume absorption coefficient in model atmosphere work, so κ rather than k is used in the present chapter.) Thus, the equations which describe the physical state of the atmosphere must be solved along with the equation of transfer, and the problem is much more complicated. A listing of the expected values of the important physical quantities as functions of depth in an atmosphere is known as a model atmosphere. In the general case one must construct a model atmosphere in order to find the emitted radiation field.

The simultaneous solution for the model atmosphere and the radiation field is very complicated, and the usual procedure is to perform an iteration. As will be shown later, far from the boundary the solution simplifies to the same form as the gray problem, and an iteration is not necessary; however, the atmosphere is the region where, by definition, one is near the boundary. One first constructs a trial model atmosphere and then tests it for consistency with its radiation field. The computed departures from radiative equilibrium then serve as clues to the modification of the trial atmosphere. Most of the concern in the present section is with the question of the construction of the model atmosphere itself.

For most stellar atmospheres hydrostatic equilibrium is a very good approximation. This means there are no large unbalanced forces on the material, the pressures pushing upward balance the gravity acting downward. If P is the total pressure, x the linear distance measured into the star, ρ the mass density and g the acceleration of gravity, then hydrostatic equilibrium requires

$$dP = g\rho \, dx. \tag{11.1}$$

If the star is rapidly rotating, g must include the effects of the centrifugal acceleration, and it will vary with position on the star. Otherwise g includes only gravitational forces, and for a thin atmosphere it is given by its surface value:

$$g = \frac{GM}{R^2} . \tag{11.2}$$

M and R are the mass and radius of the star and G is the gravitational constant. For the relatively few stars having thick atmospheres, the variation of g with depth should be taken into account; for these stars, however, curvature of the atmosphere is important and the plane approximation should not be made.

The pressure P includes both the gas pressure P_g and the radiation pressure P_r. If convective motions are present, the hydrodynamic pressure $\rho v^2/2$ may also be included, although this term is usually very small. Similarly the magnetic pressure $H^2/8\pi$ may be important in certain stars or in certain regions, such as sunspots, where the magnetic field is very large.

If only gas and radiation pressures are important, equation (11.1) is

$$\frac{dP_g}{dx} + \frac{dP_r}{dx} = g\rho . \tag{11.3}$$

Since P_r depends directly on the radiation field, the problem is more involved if radiation pressure is important. Equation (6.13) shows that

$$\frac{dP_{r_\nu}}{dx} = \frac{\pi\kappa\rho}{c} F_\nu , \tag{11.4}$$

so equation (11.3) can be written

$$\frac{dP_g}{dx} = \rho \left(g - \frac{\pi}{c}\int_0^\infty \kappa F_\nu \, d\nu\right). \tag{11.5}$$

It is possible under some circumstances to make a simplification of the above. Define a mean absorption coefficient or opacity by

$$\kappa_o = \frac{1}{F}\int_0^\infty \kappa F_\nu \, d\nu. \tag{11.6}$$

Then equation (11.5) is

$$\frac{dP_g}{dx} = \rho \left(g - \frac{\pi\kappa_o F}{c}\right) . \tag{11.7}$$

If one introduces a mean optical depth $d\tau_0 = \kappa_0 \rho dx$ and replaces F with T_e, one has

$$\frac{dP_g}{d\tau_0} = \frac{g}{\kappa_0} - \frac{\sigma T_e^4}{c} . \tag{11.8}$$

It seems that one has not gained by introducing the opacity through equation (11.6), as κ_0 depends on F_ν which is not known until the complete solution has been obtained. It is shown in the next section, however, that one can find an opacity which approximately satisfies equation (11.6) without having prior knowledge of F_ν. If this approximation is sufficient, then equation (11.8) can be used directly. If not, then equation (11.5) or its equivalent must be used in an iterative procedure involving the second term on the right side.

Radiation pressure may be neglected for all except the very hottest stars. This can be seen from a very rough calculation. If LTE is not too bad an approximation, the radiation pressure characteristic of the atmosphere of a star can be found from equation (1.10) by replacing the intensity with the Planck function of the effective temperature:

$$P_r \simeq \frac{4\pi}{3c} B(T_e) = \frac{4\sigma}{3c} T_e^4 . \tag{11.9}$$

If radiation pressure is small, equation (11.8) indicates that the typical gas pressure in an atmosphere is about

$$P_g = g\int_0^1 \frac{d\tau_0}{\kappa_0} \simeq \frac{g}{\kappa_0} , \tag{11.10}$$

where in the final expression in (11.10) the opacity is a typical value for the atmosphere as a whole. In this very rough calculation the atomic cross sections can be assumed to have their typical value of about 10^{-24} cm^2 per particle. Since a typical free particle is a hydrogen atom of mass about 10^{-24} g, the mass absorption coefficient κ_0 is about 1 cm^2 g^{-1}. One thus finds that the gas pressure in a stellar atmosphere is of the order of numerical size of the surface gravity if cgs units are used. From equation (11.9) one has

$$\frac{P_r}{P_g} \simeq 10^{-15} \frac{T_e^4}{g} , \tag{11.11}$$

where cgs units are used. Most stars fall along what is known as the main sequence and have surface gravities of the order of $g \simeq 10^4$ cm sec^{-2} For such stars, relation (11.11) predicts that radiation will make a ten per cent contribution to the gas pressure if $T_e = 3\times10^4$ °K. More accurate calculations substantially agree with this conclusion. For the remainder of this work, the analysis will be intended for the overwhelming majority of the stars for which P_r is negligible, unless explicitly stated otherwise.

With $P_g = P$, the total pressure, there is no longer any concern over whether equation (11.6) is satisfied. Instead let κ_0 be an opacity which is related to the absorption coefficient κ in a manner which is yet to be specified. Then the hydrostatic equilibrium relation (11.8) is

$$\frac{dP}{d\tau_o} = \frac{g}{\kappa_o} . \tag{11.12}$$

Let N_{ijn} be the number of particles per unit mass of the stellar material of element i, in ionization stage j, and in excitation level n. Let a_{ijn} be the atomic absorption coefficient in cross sectional area per particle of type i, j, and n. a_{ijn} will depend on the frequency of the radiation. Then in analogy with equation (2.6), the total mass absorption coefficient at a given frequency is

$$\kappa = \sum_{ijn} N_{ijn} a_{ijn} . \tag{11.13}$$

The cross section a_{ijn} depends on the properties of the different energy levels as well as on the frequency. N_{ijn} depends on the chemical abundances of the elements and on the distribution of the elements over the different ionization and excitation states, the latter being determined by temperature and electron pressure if the assumption of LTE is appropriate. Then equation (11.13) can be expressed in a schematic form by the relation,

$$\kappa = \kappa(\nu, z_i, T, P_e) . \tag{11.14}$$

z_i is the abundance of the ith element. The opacity κ_o depends on κ in some fashion, so a relation similar to equation (11.14) could be given for it also.

If x_i is the average number of free electrons supplied by the nucleus of type i, then the electron pressure is related to the total pressure by

$$\frac{P_e}{P} = \frac{\sum_i z_i x_i}{\sum_i z_i (1 + x_i)} . \tag{11.15}$$

It is assumed that the matter behaves as a perfect gas. In equation (11.15), z_i is the abundance of element i by number of atoms. The x_i's can be found from the ionization relations as functions of T and P_e. This enables equation (11.15) to be represented in the schematic form

$$P_e = P_e(z_i, P, T) . \tag{11.16}$$

Through this relation, the independent variable in equation (11.14) can be changed from P_e to P:

$$\kappa = \kappa(\nu, z_i, T, P) . \tag{11.17}$$

The meaning of the expressions (11.14), (11.16), and (11.17) is that the numerical value of the quantity (absorption coefficient or electron pressure) can be calculated, at least in principle, if the numerical values of the arguments are specified. In practice these relations are usually provided in the form of rather extensive numerical tables.

Assume for the moment that the relation between temperature and optical depth τ_o is known, where as usual $d\tau_o = \kappa_o \rho\, dx$. Then if the abundances z_i are given, the opacity κ_o depends only on P and τ_o (The opacity does not depend upon frequency.), as T can be replaced by τ_o in the realtion analogous to equation (11.17). But P and τ_o are the integration variables in the hydrostatic equilibrium relation (11.12), so this can be integrated immediately. One thus is able to find T, P, and κ_o as functions of optical depth τ_o. Through equation (11.15) or equation (11.16) P_e becomes known at each depth, while equation (11.14) yields the absorption coefficient for any desired frequency. The abundances and ionization conditions are sufficient to determine m, the average mass per free particle. The density follows from the perfect gas equation of state:

$$P = \frac{k}{m} \rho T. \tag{11.18}$$

(Here k is the Boltzmann constant.) It is seen that the initial assumption of the temperature-optical depth relation is sufficient to determine the complete model atmosphere.

The big question, of course, is "How does one find the T-τ_o relation?" The answer is "By any means available." Actually, a great deal of effort has been applied to the two problems of obtaining a good first approximation to the temperature-optical depth relation and improving this first approximation when it is not sufficient. These are the subjects of the next two sections.

A given T-τ_o relation, then, is the basis of a trial model atmosphere. The test of this model is whether it is consistent with radiative equilibrium, that is, whether the radiative flux is constant with depth. For any frequency one can calculate κ and τ at each depth. Thus the monochromatic flux F_ν can be found throughout the atmosphere. When this has been carried out for enough frequencies, the integrated flux F can be computed and the radiative equilibrium test made. The amount of work involved in this testing is quite large, but it is straightforward. Modern computers can be programmed to carry this out in a matter of minutes.

It will be noted that three different quantities had to be known or assumed before the above procedure could be carried out: T_e (or F), g, and the chemical abundances z_i. These three quantities, the so-called parameters of a model stellar atmosphere, uniquely determine the structure of the model stellar atmosphere, within the framework of the physical theory used. If this physical theory is accurate, then the model atmospheres should show in detail the circumstances of real stellar atmospheres. If the theory is not accurate, then the models will not be good numerical representatives of real stars; however, the accuracy of the physical theory, including the assumption of LTE, radiative equilibrium, plane geometry, accurately known absorption cross sections, etc., does not affect the general conclusion that the entire structure of a stellar atmosphere, including its emitted radiation field, is uniquely determined by the three parameters T_e, g, and z_i. Exceptions to the latter statement are provided only by effects not considered at all, including such phenomenon as the presence of large magnetic fields, rapid rotation, and phenomena in close binary systems.

12. Reduction to the Gray Solution

One of the methods for trying to obtain a good guess for $T(\tau_0)$ consists of trying to reduce the non-gray problem so that it has the gray solution. The fact that real stars of moderate temperatures look surprisingly accurately like gray bodies gives a basis for expecting some degree of success with this method.

The opacity κ_0 still has not been specified. (It is still assumed that P_r is very small, so that one need not be concerned with equation (11.6).) The question is whether κ_0 can be defined in such a way that $T(\tau_0)$ will be the same function as $T_g(\tau)$, the gray temperature distribution. $T_g(\tau)$ is known, being given by equation (8.11) in the Eddington approximation and by equation (9.22) in higher approximations; thus the desired definition of κ_0 would make $T(\tau_0)$ known, and the procedure of the last section could then be applied.

The equation of transfer at frequency ν is

$$\frac{\mu}{\kappa\rho}\frac{dI_\nu}{dx} = I_\nu - B_\nu , \qquad (12.1)$$

while it is desired that the opacity be such that it satisfies the gray equation:

$$\frac{\mu}{\kappa_0\rho}\frac{dI}{dx} = I - B. \qquad (12.2)$$

One can arrive directly at equation (12.2) from equation (12.1) by making the opacity satisfy

$$\frac{1}{\kappa_0}\frac{dI}{dx} = \int_0^\infty \frac{1}{\kappa}\frac{dI_\nu}{dx}\,d\nu . \qquad (12.3)$$

This equation cannot be satisfied rigorously because the intensity depends upon direction, while this cannot hold for the opacity if the analysis of the last section is to be valid. In addition the intensity is one of the unknowns of the problem, so equation (12.3) would not help in finding values of κ_0 from which the solution is to be found.

But the intensity becomes nearly isotropic in the deeper layers of the atmosphere, so a definition of κ_0 which is based on equation (12.3) is a possibility. If equation (12.1) is multiplied by κ/κ_0 and integrated over frequency, the result is

$$\frac{\mu}{\kappa_0\rho}\frac{dI}{dx} = \int_0^\infty \frac{\kappa}{\kappa_0}(I_\nu - B_\nu)\,d\nu$$

$$= I - B + \int_0^\infty \left(\frac{\kappa}{\kappa_0} - 1\right)(I_\nu - B_\nu)\,d\nu . \qquad (12.4)$$

If one could make the last term in equation (12.4) vanish, then the problem would reduce to the gray problem. Since this cannot be done, in general, one would like to make this term as small as possible:

$$\left| \int_0^\infty \left(\frac{\kappa}{\kappa_o} - 1\right)(I_\nu - B_\nu)\, d\nu \right| << \left| I - B \right|. \tag{12.5}$$

One might think of trying to satisfy equation (12.5) on the average, replacing the intensity by the mean intensity; however, this inequality could never be satisfied in radiative equilibrium, as can be seen from equation (6.12).

The equation of transfer can be written in the following way:

$$I_\nu = B_\nu + \mu \frac{dI_\nu}{d\tau},$$

$$= B_\nu + \mu \frac{d}{d\tau}\left(B_\nu + \mu \frac{dI_\nu}{d\tau}\right),$$

or

$$I_\nu = B_\nu + \mu \frac{dB_\nu}{d\tau} + \mu^2 \frac{d^2 B_\nu}{d\tau^2} + \ldots . \tag{12.6}$$

The first term on the right side of equation (12.6) is isotropic, while all of the others depend upon direction. Since the intensity tends to become isotropic far from the boundary, it follows that the first term must dominate the others at great depths in the atmosphere. The intensity approaches the Planck function.

If one uses only the first two terms in the expansion (12.6), the opacity can be defined so that the left side of equation (12.5) actually is zero. One has

$$\int_0^\infty \left(\frac{\kappa}{\kappa_o} - 1\right)\mu \frac{d3_\nu}{d\tau}\, d\nu = 0 = \frac{\mu}{\kappa_o \rho}\int_0^\infty \frac{dB_\nu}{dx}\, d\nu - \frac{\mu}{\rho}\int_0^\infty \frac{1}{\kappa}\frac{dB_\nu}{dx}\, d\nu.$$

dB_ν/dx is not known, as it depends on the structure of the atmosphere; however, the derivative can be replaced by $(dT/dx)(dB_\nu/dT)$, and the first factor cancels because it does not depend on the frequency. The above then leads to the following expression for the opacity:

$$\frac{1}{\kappa_o}\frac{dB}{dT} = \int_0^\infty \frac{1}{\kappa}\frac{dB_\nu}{dT}\, d\nu. \tag{12.7}$$

The opacity defined by equation (12.7) is known as the Rosseland mean absorption coefficient, and it is obviously analogous to equation (12.3).

If one is not too close to the surface, the use of the optical depth defined in terms of the Rosseland mean causes the $T(\tau_o)$ relation to be the same numerical function as $T_g(\tau)$, the known grey temperature distribution. This is the reason the Rosseland mean is universally used in stellar interior work, as the approximations leading to equation (12.7) are extremely accurate. An atmosphere is close to the surface, and the first two terms of equation (12.6) are not a very accurate representation of the intensity; nevertheless, the use of the Rosseland mean allows one to obtain a reasonalby good first approximation to the correct temperature distribution. This approximation is sufficient for many purposes, but if higher accuracy is needed, a method for correcting a given distribution is used. Such methods of correction are examined in the next section.

If the expansion (12.6) is used to evaluate the flux, it is seen that only terms with odd powers of μ make a contribution. The two (or three) term expansion yields

$$F_\nu = \frac{4\pi}{3}\frac{dB_\nu}{d\tau} = \frac{4\pi}{3\kappa_0\rho}\frac{dB_\nu}{dx} = \frac{4}{3}\frac{\pi}{\kappa_0\rho}\frac{dT}{dx}\frac{dB_\nu}{dT} . \qquad (12.8)$$

The flux depends directly on the deviations of the intensity from isotropy. If dB_ν/dT from equation (12.8) is substituted into equation (12.7), one finds the following alternate expression for the Rosseland mean:

$$\kappa_0 = \frac{1}{F}\int_0^\infty \kappa F_\nu \, d\nu. \qquad (12.9)$$

This is identical to equation (11.6); thus, the Rosseland mean also allows the radiation pressure to be approximately expressed in the simple form indicated in equation (11.8).

Other opacities have been introduced in order to improve upon the Rosseland mean for stellar atmospheres. The most important of these is the Chandrasekhar mean, given by an expression similar to equation (12.9) except that gray body fluxes are used instead of the actual fluxes. The Chandrasekhar mean is designed specifically for the atmosphere, but it rests on certain approximations whose accuracy is impossible to evaluate in the general case. In the specific cases in which it has been applied, it seems to be a slight improvement on the Rosseland mean in accuracy, but it also makes the calculations more difficult. In view of the efficient methods now available to correct temperature distributions, it is perhaps best to simply use the absorption coefficient κ at some given frequency for the opacity.

13. Corrections to the Temperature Distribution

If the use of the appropriate opacity does not yield a model atmosphere which reproduces the desired flux condition to sufficient accuracy, then one must resort to methods which correct a given model. One has calculated the fluxes of the trial model, so one knows the magnitude of the error in the fluxes. The problem is to find the necessary change in temperature at each depth to correct the fluxes by those known amounts.

The problem is complicated by the fact that the flux is not fixed by conditions at the depth in question, but it depends on conditions both above and below the given point. Thus, suppose that the temperature of a thin layer at a given depth is increased by a certain amount, other layers being unchanged. Then the energy output of the given layer is increased, and this causes all higher layers to have a larger upward flux, while all lower layers will have a smaller upward flux. The amounts of these changes depend on the optical distance from the layer in question as well as on the optical thickness of the region in which the temperature is changed. All of these optical distances depend on the frequency variation of the absorption coefficient κ. The problem is further complicated by the fact that most of the simplifications which one might think of are valid only at large depths, and so are of little help in the higher regions of the atmosphere.

A very large number of papers have been published on methods for correcting a trial model atmosphere. Most of these methods depend in some way on the assumption that departures from grayness are not large, and they are thus rather restricted in their application. One of the more useful methods, not so restricted, is due to E.H. Avrett and Max Krook (*Ap. J.* **137**, 874, 1963), and the remainder of the present section is based on this method. For a discussion of other methods and the relevant references, one could see J.C. Pecker's article in *Ann. Rev. Astron. Astrophys.* **3**, 135, 1965. The remainder of this section contains mathematical analysis with relatively little physical content relevant to an understanding of stellar atmospheres. It is intended for those who might want more detailed knowledge on the practical aspects of building model stellar atmospheres.

If n is the ratio κ/κ_0, where κ_0 is defined in any way that is convenient (not necessarily the Rosseland mean), then the equation of transfer can be written

$$\mu \frac{dI}{d\tau_0} = n(I_\nu - B_\nu). \tag{13.1}$$

All quantities are now expressed as the sum of a zero order term (subscript 0) and a first order term (subscript 1):

$$\begin{aligned} \tau_0 &= t_0 + t_1, \\ I_\nu(\tau_0,\mu) &= I_{0\nu}(t_0,\mu) + I_{1\nu}(t_0,\mu), \\ T(\tau_0) &= T_0(t_0) + T_1(t_0). \end{aligned} \tag{13.2}$$

Any of the other variables can be written in a similar fashion. The first order term for n is written in the form $t_1 n'$, where the prime stands for the derivative with respect to t_0; the first order term for B_ν is likewise $T_1 dB_\nu/dT$. If this expansion is made for the quantities appearing in equation (13.1), and if the result is separated into zero and first order parts, then the following two equations are obtained:

$$\mu \frac{dI_{0\nu}}{dt_0} = n(I_{0\nu} - B_{0\nu}) \tag{13.3}$$

$$\mu \frac{dI_{1\nu}}{dt_0} = n\left(I_{1\nu} - T_1 \frac{dB_\nu}{dT}\right) + (nt_1' + t_1 n')(I_{0\nu} - B_{0\nu}). \tag{13.4}$$

In the above, products of two first order terms are neglected. If these equations are multiplied by μ^m and integrated over the solid angle, the following relations between the first three moments of the intensity are obtained for $m = 0, 1$:

$$\frac{dF_{0\nu}}{dt_0} = 4n(J_{0\nu} - B_{0\nu}), \tag{13.5}$$

$$\frac{c}{\pi}\frac{dP_{r0\nu}}{dt_0} = nF_{0\nu} , \tag{13.6}$$

$$\frac{dF_{1\nu}}{dt_0} = 4n\left(J_{1\nu} - T_1\frac{dB_\nu}{dT_0}\right) + 4(nt_1' + n't_1)(J_{0\nu} - B_{0\nu}), \tag{13.7}$$

$$\frac{c}{\pi}\frac{dP_{r1\nu}}{dt_0} = nF_{1\nu} + (nt_1' + n't_1)F_{0\nu} \tag{13.8}$$

The problem is to find the new temperature-optical depth relation such that the flux condition is more accurately satisfied. The only approximation made so far is that the first order terms are small enough that products of them can be neglected. t_1 is as yet completely arbitrary and was introduced to facilitate the solution for T_1.

Solve equation (13.7) for T_1 and integrate it over frequency. Noting that $F_1 = F - F_0$ and eliminating $(J_{0\nu} - B_{0\nu})$ through equation (13.5), one finds

$$T_1\int_0^\infty n\frac{dB_\nu}{dT}\,d\nu = \int_0^\infty nJ_{1\nu}\,d\nu + \frac{t_1}{4}\int_0^\infty \frac{n'}{n}\frac{dF_{0\nu}}{dt_0}\,d\nu - \frac{1}{4}\frac{dF}{dt_0}$$
$$+ \frac{1}{4}(1 + t_1')\frac{dF_0}{dt_0} . \tag{13.9}$$

If an original temperature distribution is known or assumed, that is, if $T_0(t_0)$ is known, then all of the zero order quantities can be found by the methods outlined in Section 11. The desired flux F is known (being equal to $\sigma T_e^4/\pi$ for radiative equilibrium), so dF/dt_0 is also known. t_1 can be chosen arbitrarily, so all quantities in equation (13.9) are known except T_1 and $J_{1\nu}$. If the latter can somehow be determined, then T_1, the temperature correction, follows. An additional approximation has to be made in order for $J_{1\nu}$ to be found, and t_1 will be defined so as to make this a reasonably good one.

Now t_1 will be defined to be the solution of the following differential equation:

$$t_1' F_0 + t_1\int_0^\infty \frac{n'}{n}F_{0\nu}\,d\nu + F - F_0 = 0. \tag{13.10}$$

The boundary condition is that $t_1(0) = 0$. t_1 can thus be determined from zero order quantities. In view of the above, equation (13.8) becomes

$$\int_0^\infty \frac{1}{n}\frac{dP_{r1\nu}}{dt_0}\,d\nu = 0. \tag{13.11}$$

In the Eddington approximation $P_{r\nu} = 4\pi J_\nu/3c$, as in equation (8.4), so

$$\int_0^\infty \frac{1}{n}\frac{dJ_{1\nu}}{dt_0}\,d\nu = 0. \tag{13.12}$$

A sufficient, though not necessary, condition for the validity of equation (13.12) is

$$\frac{dJ_{1\nu}}{dt_0} = 0, \tag{13.13}$$

for all frequencies. Equation (13.13) means that $J_{1\nu}$ is independent of depth, so it equals its surface value. In the Eddington approximation $J_\nu(0) = \frac{1}{2}F_\nu(0)$ as in equation (8.6), so

$$2J_{1\nu} = F_{1\nu}(0) = F_\nu(0) - F_{0\nu}(0). \tag{13.14}$$

One often sees the coefficient of $J_{1\nu}$ in the above as $4\sqrt{3}/3$ instead of 2, but the difference is far below the present level of approximation.

Equation (13.14) still does not give $J_{1\nu}$ in terms of known quantities, as the final monochromatic flux at the surface $F_\nu(0)$ is not known. From equations (13.8) and (13.13) along with the boundary condition on t_1, however, one finds that

$$F_{1\nu}(0) + t_1'(0)\, F_{0\nu}(0) = 0,$$

or equivalently

$$F_\nu(0) = [1 - t_1'(0)]\, F_{0\nu}(0) = \frac{F(0)F_{0\nu}(0)}{F_0(0)}. \tag{13.15}$$

Equations (13.14) and (13.15) finally fix $J_{1\nu}$ in terms of known quantities, and the expression for T_1 becomes

$$T_1\int_0^\infty n\,\frac{dB_\nu}{dT}\,d\nu = \frac{F(0) - F_0(0)}{2F_0(0)}\int_0^\infty nF_{0\nu}(0)\,d\nu + \frac{t_1}{4}\int_0^\infty \frac{n'}{n}\,\frac{dF_{0\nu}}{dt_0}\,d\nu$$

$$+ \frac{1}{4}(1 + t_1')\,\frac{dF_0}{dt_0} - \frac{1}{4}\,\frac{dF}{dt_0}. \tag{13.16}$$

Equations (13.10) and (13.16) are sufficient to fix the new temperature distribution in the original trial model atmosphere. Several applications of this procedure may be required to bring the errors in the calculated fluxes down to those which can be tolerated.

Only a series of numerical tests can tell whether the various approximations made in the above analysis are sufficient for the method to converge rapidly to the correct solution. Practical experience indicates that the Avrett-Krook method is very useful; the very large number of calculations necessary make it out of the question for hand calculations, but a large computer can finish a correction in a matter of seconds, and the convergence toward the correct fluxes is usually very rapid. In fact the method can be considerably simplified by ignoring the frequency dependence of n, and the convergence is still very good in most practical cases, as indicated in *Ap. J.* **141**, 821, 1965. Thus one is able to construct a model atmosphere which satisfies a given flux condition to an arbitrary degree of accuracy.

14. Convection

The combination of plane geometry and the assumption that the nuclear sources are deep within the stars requires that the total energy flux in the outward direction be constant throughout the atmosphere. Previously, it had been assumed that radiation is the only significant means of transporting the energy out of a star, so a radiative flux constant with depth was the result; however, if convection occurs which is also carrying some of the energy, then it is the sum of the radiative and convective fluxes which is constant with depth, and the previous methods of constructing a model atmosphere need to be modified. When convection is important, radiative equilibrium no longer holds.

If πF_r and πF_c are the radiative and convective fluxes, then

$$F_r + F_c = \frac{\sigma}{\pi} T_e^4 . \tag{14.1}$$

F_r is calculated as before, but now F_c also needs to be calculated.

The presence of convection means, of course, that there are large scale mass motions. Thus hydrostatic equilibrium cannot be exactly realized, and the equations of hydrodynamics may have to be considered; however, the motions are usually slow enough that hydrostatics remains a very good approximation. In the solar atmosphere for example, since the gas pressure is about 10^5 cgs where the mass density is 10^{-7} cgs, and large scale velocities v (from observations) are of the order of a kilometer per second or 10^5 cgs, the hydrodynamic pressure $\rho v^2/2$ is no more than about one percent of the gas pressure. Thus for purposes of further discussion and without much loss of physical reality, hydrostatic equilibrium is assumed.

One needs to determine first the conditions under which convection will occur. In an atmosphere consider a mass element in equilibrium with its surroundings, that is, the same temperature, pressure, and density. Suppose that a random perturbation causes this element to be displaced downward by a small amount. The element is in new surroundings with different physical conditions, and has a force exerted on it as a result. If the force is upward, tending to restore the element to its original position, the material is stable against convection: motions are damped out, and convection does not occur. If the force is downward, tending to push the element farther from its original position, the material is unstable against convection: random motions are enhanced, and convection occurs.

Let P, T and ρ be the original pressure, temperature, and density of the element, respectively. If δx is the distance the element is displaced downward, then the new density of the surroundings is $\rho + \delta x(d\rho/dx)_{atm}$, where the subscript atm means the given quantity is to be evaluated at the given point in the atmosphere. If one assumes that the element immediately contracts until it has the same pressure as the new surroundings, and if this contraction is rapid enough to be essentially adiabatic, then the new density of the element will be $\rho + \delta P(d\rho/dP)_{ad}$, where $\delta P = \delta x(dP/dx)_{atm}$. The subscript ad means the quantity is evaluated for an adiabatic change. The mass element sinks further if it is more dense than its new surroundings, so the condition for convection to occur is

$$\left.\frac{d\rho}{dP}\right)_{ad} > \left.\frac{d\rho}{dP}\right)_{atm} . \tag{14.2}$$

If the density in the atmosphere increases less rapidly than the adiabatic gradient, the lower layers are not able to support the upper ones without becoming unstable and allowing convective motions. The perfect gas equation of state (11.18) allows the condition (14.2) to be expressed in terms of temperature instead of density:

$$\frac{d}{dP}\left(\frac{T}{m}\right)_{atm} > \frac{d}{dP}\left(\frac{T}{m}\right)_{ad} . \tag{14.3}$$

Here m is the average mass per free particle. If m is nearly a constant, then inequality (14.3) takes the more familiar form,

$$\left.\frac{dT}{dP}\right)_{atm} > \left.\frac{dT}{dP}\right)_{ad} , \tag{14.4}$$

as the condition that convection exist.

If the temperature gradient at a given depth is superadiabatic, convection occurs. In constructing a model atmosphere, one must calculate the adiabatic gradient in order to make the convection test. For a perfect gas which is completely neutral or completely ionized, $dT/dP)_{ad} = 0.4T/P$. When the material is partly ionized, the adiabatic gradient is a complicated function of physical conditions. For a derivation of this gradient under more general conditions, one can see, for example, Unsöld's book *Physik der Sternatmosphären*.

When one constructs a model atmosphere without the restriction of radiative equilibrium, one must check at each point to see if the inequality (14.4) is satisfied. If the inequality is not satisfied, $F_c = 0$ at that point. (Actually there can be some overshooting of convection into the neighboring region.) If inequality (14.4) is satisfied, then one must be able to calculate F_c so that condition (14.1) can be tested.

The convective flux is generally calculated by the highly idealized mixing length theory. It is supposed that the convective medium is composed of a large number of small mass elements, each of which moves as a unit. The elements are continually being formed, they move (up or down) a certain distance, and then dissipate back into the background medium. The average vertical distance they travel before losing their identity is known as the mixing length L. There is an average value of the temperature at each depth which is somewhat cooler than the rising elements and somewhat hotter than the falling ones.

Let T, P, and ρ be the values of the variables on the average at a given depth. Let $T + \delta T$, $\rho + \delta\rho$ be the temperature and density of the typical elements at the same depth. For rising elements δT will be positive and $\delta\rho$ negative, while for falling ones these will be reversed. It is generally assumed that the elements have expanded or contracted until they are at the same pressure P as the surroundings. If c_p is the specific heat per unit mass at constant pressure, then $c_p\,\rho\,\delta T$ is the excess energy per unit volume in the elements (positive for rising elements and negative for falling ones). Thus the vertical motion of the mass elements causes a net upward flux of energy. If v is the mean velocity of the convective elements, measured positive upward in the direction of the flux, then

$$\pi F_c = c_p\,\rho\,v\,\delta T. \tag{14.5}$$

The velocity and the temperature difference must now be found in terms of local variables.

The buoyant force exerted on an element per unit volume is $g\delta\rho$, where g is the acceleration of gravity. The work done on the element, W, is then

$$W = \int_0^{L/2} \delta\rho \; g dx \; . \tag{14.6}$$

L is the average distance a convective element travels before it disappears, so an element chosen at random is accelerated over a distance of roughly L/2 (hence the upper limit in the integral). If $\delta\rho$ is taken to be proportional to the displacement for convenience (the final result is not sensitive to this assumption), the above becomes

$$W = \frac{1}{4} gL \; \delta\rho \, . \tag{14.7}$$

Since the elements have the same pressure as the surroundings, $\delta\rho = -\rho\,\delta T/T$ for a perfect gas, and equation (14.7) becomes

$$W = \frac{gL\rho\delta T}{4T} = \frac{gL^2\rho}{8T} \left| \left.\frac{dT}{dx}\right)_{el} - \left.\frac{dT}{dx}\right)_{atm} \right| . \tag{14.8}$$

The subscript el indicates that the quantity is determined within the mass element as it moves. δT in (14.8) is also evaluated at a distance of L/2. W is the average buoyant energy per unit volume received by an element chosen at random; if this is equated to the kinetic energy per unit volume $\rho v^2/2$, one obtains for the velocity

$$v^2 = \frac{gL^2}{4T} \left| \left.\frac{dT}{dx}\right)_{el} - \left.\frac{dT}{dx}\right)_{atm} \right| . \tag{14.9}$$

If there is no energy exchange between the element and the surroundings during the lifetime of the element, then the temperature gradient within the element is the adiabatic gradient. One finally obtains for the flux:

$$F_c = c_p\rho\left(\frac{g}{T}\right)^{1/2} \frac{L^2}{4} \left| \left.\frac{dT}{dx}\right)_{atm} - \left.\frac{dT}{dx}\right)_{ad} \right|^{3/2} . \tag{14.10}$$

The absolute value bars are indicated in this to emphasize that a positive difference in temperature gradients is needed, although in the present situation the coordinate system is such that they could be removed from equation (14.10).

The mixing length theory does not determine the value of L, so it is a free parameter. One might expect it to be of the order of the scale height in the atmosphere, perhaps, as the elements should lose their identity if they move into conditions very different from where they were created. Another possibility is that L is of the order of the distance to the nearest boundary of the convective region, if the latter is smaller than the scale height. As the theory leading to equation (14.10) is somewhat artificial, one should not be surprised if the numerical coefficient needs to be adjusted by several factors of two. Different possible values of L are due to undertainties in the convective theory. L is not an independent parameter of the atmosphere in the sense of T_e, g, and the chemical composition.

A more sophisticated analysis is possible, for example, by taking into account ionization changes in the relation between $\delta\rho$ and δT, and by calculating the energy exchange between element and surroundings, instead of assuming the temperature gradient,for the element is the adiabatic gradient. For more detailed treatments see E. Vitense, *Zs. f. Ap.* **32**, 135,

1953, and E. Böhm-Vitense, *Zs. f. Ap.* **46**, 108, 1958. The framework of the mixing length theory, however, probably places a limit on the accuracy with which the actual convection in stars can be represented by relations such as equation (14.10).

Stars which are not too hot have a convection zone in the lower part of their atmospheres. The properties of the zone are strongly influenced by the excitation and the ionization of hydrogen, so it is called the hydrogen convection zone. Its upper boundary is near optical depth unity, its precise position depending on the type of star. The cause of this outer convection region can be understood from the following very rough analysis:

Suppose that one has a gray atmosphere. If convection is not important, the temperature distribution will be given approximately by equation (8.11):

$$T^4 = \frac{3}{4} T_e^4 \left(\tau + \frac{2}{3}\right) . \tag{14.11}$$

Taking the derivative of this expression and using the equation of hydrostatic equilibrium (11.12), one finds

$$\frac{P}{T}\frac{dT}{dP} = \frac{\kappa P}{4g(\tau + 2/3)} . \tag{14.12}$$

Convection will occur if this is greater than the corresponding adiabatic value, which is 0.4 if ionization is not changing with depth. At the surface both P and τ go to zero, and these layers are not convective. As one considers deeper layers, the physical conditions change and the question of the existence of convection depends on numerical details.

Suppose that the absorption coefficient κ is proportional to P^n. This is approximately true over a limited range of depths, and it illustrates the main cause of the hydrogen convection zone. For large positive n, the absorption increases rapidly with depth, while for large negative values it drops off rapidly with depth. For this assumed form and the same boundary condition on P and τ as above, integration of the hydrostatic equilibrium relation gives $\kappa P = g\tau(n + 1)$, so convection occurs if

$$\frac{\tau}{\tau + 2/3} > \frac{1.6}{n + 1} . \tag{14.13}$$

This relation cannot be satisfied for any τ unless n is somewhat greater than +0.6. The absorption must increase with depth faster than this minimum rate in order for a convection zone to occur. The more rapid the increase of absorption with depth, the closer to the surface is the upper boundary. For the upper boundary to be near $\tau = 1$, n must be close to 1.6. The assumption that the absorption is proportional to P^n does not imply such a physical dependence, but that both P and κ change with depth in such a way that this is an adequate interpolation formula.

When F_c is quite significant, then radiative equilibrium is not valid. In such a case, if radiative equilibrium is used as the basis of the first approximation, the trial atmosphere needs a considerable modification.

In the atmospheres of moderate temperature stars, the dominant source of absorption is the negative ion of hydrogen, H^-. The formation of this ion requires the collision of a free electron and a neutral hydrogen atom, so that the abundance of H^- is very sensitive to the density. As a result, κ increases rapidly with depth. A secondary cause is that, at lower temperatures, the $n = 2$ level of hydrogen has a population which is very sensitive to temperature. In the deeper layers, where H is becoming important, κ will again have a strong increase with depth. It might also be mentioned that the ionization of hydrogen lowers the adiabatic temperature gradient somewhat, and equation (14.4) indicates that this, too, favors convection.

The occurrence of convection does not necessarily rule out radiative equilibrium. The very low densities in stellar atmospheres make convection an inefficient carrier of energy. The temperature gradient can be considerably greater than the adiabat without F_c being appreciable. In the deeper layers convection becomes very efficient, and only a very small difference between the actual gradient and the adiabat is sufficient to carry essentially all of the flux; however, this region occurs far below the atmosphere.

Numerical calculations and observations show that convection has only a small effect on the atmospheric structure and emitted radiation of most stars. Unless there is particular interest in the relatively deep layers of an atmosphere, one can retain the assumption of radiative equilibrium for most types of stars.

15. Semi-Empirical Models

The methods described previously for constructing a model atmosphere are entirely theoretical, but a method exists which uses observations directly. Unfortunately, the detailed observations necessary are available only for the Sun, so the method has a decided limitation; however, it does provide an important check on the purely theoretical models.

For a semi-infinite, plane atmosphere in LTE, the intensity at the surface is

$$I_\nu(0,\mu) = \int_0^\infty B_\nu(\tau) e^{-\tau/\mu} \frac{d\tau}{\mu} \, . \tag{15.1}$$

This relation indicates that the depth dependence of the Planck function, which means the depth dependence of the temperature, is related directly to fixing the variation of intensity with μ. If $T(\tau)$ is known, then $I_\nu(0,\mu)$ is easily obtained. It is pointed out in Section 11 that $T(\tau)$ is all that is needed to construct a model atmosphere. Since $I_\nu(0,\mu)$ is observable, the problem then presents itself of finding out whether equation (15.1) can be inverted. If so, observations of $I_\nu(0,\mu)$ can be used to determine the temperature-optical depth relation needed to construct a model atmosphere.

Direct attempts at a mathematical inversion of equation (15.1) have not been particularly successful; however, by expressing the Planck function as an expansion in optical depth with adjustable parameters, and by forcing the parameters to have values which fit the observed intensity, a number of investigators have constructed semi-empirical models of the solar atmosphere. The most notable of these models is by A.K. Pierce and J.H. Waddell, *Mem. R.A.S.* **LXVIII**, 89, 1961.

Let B_ν^* be the ratio $B_\nu/I_\nu(0,1)$, where $I_\nu(0,1)$ is the intensity at the center of the disk. Expand B_ν^* as follows:

$$B_\nu^* = a + b\tau + cE_2(\tau) + \dots \quad (15.2)$$

$E_n(\tau)$ is the nth order exponential-integral function. A term by term integration of equation (15.1) indicates that the normalized intensity $I_\nu^*(0,\mu) = I_\nu(0,\mu)/I_\nu(0,1)$, called the limb darkening, is given by

$$I_\nu^*(0,\mu) = a + b\mu + c\left[1 - \mu \ln(1 + \frac{1}{\mu})\right] + \dots \quad (15.3)$$

The form of the expansion (15.2) is suggested by analogy with the gray atmosphere, but the main argument for it is that equation (15.3) gives an excellent fit to the observed intensities for the Sun. Pierce and Waddell indicate that the three term expansion fits to within the observational errors, so adding the higher order terms would not improve the accuracy.

A set of expansion parameters (a,b,c) is found by a least squares fit to the observed limb darkening at a selected frequency ν which, in turn, determines B_ν^*. Then B_ν^* is known from equation (15.2) for all depths. If the absolute intensity values at the center of the disk, $I_\nu(0,1)$, are known from observation, the Planck function and the temperature are determined as functions of τ. Enough information becomes known to construct a model atmosphere without resorting to the methods which are described in Sections 12 and 13.

There is another important property of the limb darkening observations. When one obtains the temperature-optical depth relation at a given frequency, one can differentiate it to obtain $d\tau/dT = \kappa\rho\ dx/dT$, where x is the geometric distance into the atmosphere. If one does this for two different frequencies, and if one then takes the ratio at fixed temperature, hence at fixed x and ρ, one has evaluated

$$\frac{d\tau(\nu_1)}{d\tau(\nu_2)} = \frac{\kappa(\nu_1)}{\kappa(\nu_2)} . \quad (15.4)$$

Thus one can obtain the frequency dependence of the absorption coefficient, at any depth, directly from the observations. This provides an important check on the theoretically determined absorption coefficient of the Sun. It should be noted that the semi-empirical method does not depend on the assumption of radiative equilibrium.

The semi-empirical method produces a model atmosphere which, by definition, satisfies the limb darkening observations; however, this does not mean that the model is accurate in detail. Limb darkening is known to be rather insensitive to many aspects of a model atmosphere. Thus, the three term expansion of equation (15.2) may be accurate for optical depths in the range of 0.5-1.5 or so, but it can be very inaccurate if applied to layers much outside this range. The semi-empirical model provides an important check on the solar atmosphere, but it does need supplementing by other methods. K-H. Böhm, *Ap.J.* **134**, 264, 1961, discusses the accuracy of a model based on limb darkening measures.

16. Continuous Absorption and Blanketing

In order to construct a model atmosphere, it is necessary to be able to calculate the absorption coefficient κ as a function of the chemical abundances, frequency, and physical conditions. Thus the important absorbing agents must be identified, and the absorption cross sections must be available either from calculations or from laboratory data. It is not the purpose here to reproduce either the quantum mechanical calculations of the cross sections or a long list of tables or formulas of the results. A few general comments will be made along with some references to the literature.

It has already been mentioned that the main source of continuous absorption in the atmospheres of moderate temperature stars is the negative ion of hydrogen. H^- has only one bound state about 0.75 ev below the continum; thus, photons of wavelength less than about 16,500 A can ionize H^- to produce neutral hydrogen plus a free electron. H^- never becomes very abundant: at higher temperatures it is too easily ionized, and at lower temperatures there is too small a source of free electrons. In spite of the very low abundance, the cross section for the absorption of radiation by an H^- ion is large enough for it to dominate the absorption coefficient. For wavelengths longer than 16,500 A, only free-free transitions are possible. These result from the acceleration of a free electron which is temporarily in the field of a nearby neutral hydrogen atom.

One of the interesting consequences of H^- dominating stellar absorption is that the absorption does not vary strongly with wavelength. The boundfree cross section varies by less than a factor of two between 4000A and 13,000 A, and as a result gray model atmospheres are good approximations to real atmospheres.

At other temperatures, different absorbers are important. For moderately hot stars neutral hydrogen is strong, while neutral and ionized helium, as well as electron scattering, are important for very hot stars. At lower temperatures molecules have an increasingly important effect. There are many sources of absorption which, although not very strong, are not weak enough to be neglected. There are many absorbers which are important over limited wavelength regions, but which have little or no influence on the structure of the model atmosphere. For a review of the different sources of absorption in stellar atmospheres, one can see O.J. Gingerich's article in *Smithsonian Astroph. Obs. Special Report No. 167* (First Harvard-Smithsonian Conference on Stellar Atmospheres), p. 17, 1964. Absorption coefficients from many of the metals are given by L.D. Travis and S. Matsushima, *Ap. J.* **154**, 689, 1968. References to absorption in very cool stars are found in the review article by M.S. Vardya, *Ann. Rev. Ast. Astrophys*. **8**, 87, 1970.

In addition to bound-free and free-free transitions discussed above, the bound-bound transitions also have an effect on the structure of an atmosphere and the emitted radiation field. A single line generally covers such a small wavelength interval that it has no significant effect on the energy balance of the atmosphere; however, this is not true of the total effect of thousands of lines, and a statistical way of handling all of these lines needs to be used. The cumulative effect of all lines is known as the blanketing effect, a name derived from the fact that the lines make it more difficult for radiation to escape from the atmosphere, causing an increase in temperature.

In the simplest blanketing model, the so-called picket fence representation, the absorption lines are assumed to have square profiles. The main parameter is one which fixes the relative probabilities that, at any given wavelength, one is in a line or between lines. In more sophisticated treatments, one calculates the line absorption coefficient at a given wavelength from data on the positions and shapes of a very large number of lines. It would appear that, as far as the structure of the atmosphere is concerned, the simple theories are about as accurate as the more complicated ones; however, for an accurate prediction of the frequency distribution of flux or intensity at high resolution, the more involved approaches are of course required.

Blanketing is not very well represented by simply adding a slowly varying function of wavelength to the continuous absorption coefficient. Blanketing causes a small wavelength interval to receive radiation from both high and low regions in the atmosphere, and replacing this range of contributions by a single intermediate depth of formation does not give the same effect. The proceedings of an I.A.U. conference on blanketing are contained in *J. Quan. Spec. Rad. Tran.* **6**, 539-704, 1966.

The absorption lines in stellar spectra do not occur uniformly at all wavelengths, but are much more prevalent at the shorter wavelengths. Blanketing thus distorts the radiation stars emit, causing more radiation to be emmitted at longer wavelengths than would otherwise be the case. This is a very pronounced effect, and it complicates efforts to determine information from the observed colors of stars. If the temperatures of two stars are to be compared by means of their colors, one must make sure that they have the same amount of blanketing. Since most of the blanketing is caused by lines due to the heavy elements, this means the two stars should have about the same abundances of the heavy elements. But significant variations of composition are known to exist among stars, and a correction for differences in blanketing should be made. A.R. Sandage and O.J. Eggen, *M. N. R. A. S.* **119**, 278, 1959, and R.L. Wildey, E. M. Burbidge, A.R. Sandage, and G.R. Burbidge, *Ap. J.* **135**, 94, 1962 give approximate ways of correcting observed colors for differences in blanketing.

4

LINE FORMATION

17. Line Absorption and Emission

A spectral line results when the absorbing and emitting properties of a medium change significantly over a very small frequency range. The frequency range of a line is usually small enough that the properties of the continuum can be considered as constant over the line, but there can be exceptions for very strong lines. The line represents a transition between two bound levels of an atom.

Much of the material in the earlier chapters applies equally well to line radiation and to continuum radiation. One can solve for the line radiation in a given situation when the line absorption and emission coefficients, or equivalently, line absorption coefficient and source function, are known throughout the medium. One simplification applicable to the lines but not to continuum is that, in most cases, the energy involved in a given line transition is too small to appreciably affect physical conditions in the regions where the line is formed. Thus, one can usually study the formation of a given line without the necessity of a simultaneous solution for a model atmosphere; however, the latter may become necessary if the region of line formation is not well understood from studies of the continuum or from other lines.

It is convenient to introduce Einstein coefficients for the line processes as was done for the continuum in equations (5.12). In fact the only modification needed is that the right sides of these equations should all be multiplied by the profile function ϕ_ν. $\phi_\nu \, d\nu$ is the probability that, if the given transition does take place and if the intensity is constant across the line, the absorbed or emitted photon has frequency between ν and $\nu + d\nu$. There is actually a separate profile function for absorption and for emission, but the differences between them are generally quite small (see, for example, R.N. Thomas, *Ap. J.* **125**, 260, 1957) and they are assumed identical here. The profile function is defined so that

$$\int_0^\infty \phi_\nu \, d\nu = 1. \tag{17.1}$$

The profile function is written as a separate factor for line transitions so that the Einstein coefficients can still be considered as atomic constants, independent of physical conditions.

With the above exception, the anlysis following equations (5.12) is valid for line radiation as well as continuum radiation. In particular, the absorption coefficient is given by equation (5.16) multiplied by the profile function. If a is the atomic absorption coefficient given by k/N_1, then from equation (5.16)

$$a = \left(1 - \frac{N_2 B_{21}}{N_1 B_{12}}\right) \frac{h\nu}{4\pi} B_{12} \phi_\nu \, . \tag{17.2}$$

The units of a are cm^2/atom in the lower level.

It is common to express the transition probability in terms of the so-called oscillator strength or f-value. For example, for the transition from level 1 to level 2, this is given by

$$f_{12} = \frac{mch\nu}{4\pi^2 e^2} B_{12} = \frac{g_2}{g_1} \frac{mc^3}{8\pi^2 e^2 \nu^2} A_{21} \, . \tag{17.3}$$

Equations (5.20) still hold between the Einstein coefficients. In terms of the f-value, the atomic line absorption coefficient in equation (17.2) is

$$a = \left(1 - \frac{g_1 N_2}{g_2 N_1}\right) \frac{\pi e^2}{mc} f_{12} \phi_\nu . \tag{17.4}$$

Note that ϕ_ν is the only part of the absorption coefficient that varies with frequency across the line. The first factor on the right side of equations (17.2) and (17.4) is known as the correction for induced emission. Near thermal equilibrium, equation (5.19) gives the population ratio of the levels, and this term becomes $(1 - e^{-h\nu/kT})$. The uncertainty in the value of the oscillator strength is an important source of error in line analysis, although the situation has improved much in recent years. The determination of ϕ_ν is the subject of Section 19.

Since the source function is the ratio of emission and absorption coefficients, the profile function cancels out, and equations (5.17), (5.21), and (5.23) are valid as they stand for line transitions. As before, if conditions are sufficiently close to thermal equilibrium (TE) that the excitation temperature for two given levels, as defined by equation (5.22), is very close to the kinetic temperature throughout the line forming region, then the given line is formed under the conditions of local thermal equilibrium (LTE). Lines are usually formed higher in the atmosphere than the continuum, so one can expect on quite general grounds that deviations from LTE are greater for the lines than for the continuum.

For a given atomic level j, it is common to define a departure coefficient b_j through the relation

$$N_j = b_j N_j^*, \tag{17.5}$$

where N_j^* is what the population of level j would be in TE at the same kinetic temperature, electron density, and ion density of the given element. Thus $b_j = 1$ in TE, and this also holds for continuum levels if a Maxwellian velocity distribution of the free particles prevails. The departures of the b_j's from unity are measures of departures from the Boltzmann and Saha populations. One then has

$$\frac{N_1}{N_2} = \frac{b_1}{b_2}\frac{g_1}{g_2} e^{h\nu/kT} , \tag{17.6}$$

where $\nu = E_{12}/h$ and T without a subscript represents the local kinetic temperature. If this is substituted into equation (5.21), the line source function between levels 1 and 2 is:

$$S_{12} = \frac{2h\nu^3/c^2}{(b_1/b_2)e^{h\nu/kT} - 1} . \tag{17.7}$$

The subscript ν of the line source function is suppressed for convenience. If $b_1 = b_2$, S_{12} reduces to the Planck function of the local kinetic temperature and LTE prevails for the given line. In general this will not be the case, and one must determine the ratio (b_1/b_2) throughout the line forming region. In the next section this ratio is found in the simple case of an atom having only 2 bound states, and the generalization to the many level atom is briefly described. After the profile function is discussed in Section 19, one is then in a position to solve the line formation problem.

18. The Two-Level Atom

It is indicated above that if the populations of the different atomic energy levels are the same as in TE, all of the b_j's are equal to unity, and LTE prevails in all transitions. The populations of the different levels are determined by the transition rates between the levels, and these transitions can be either collisional or radiative, i.e., the energy of the transition can be supplied by or given to either another particle or a photon.

The collisions which take place between the particles in a gas are of two kinds: elastic and inelastic. Elastic collisions exchange kinetic energy among the colliding particles, but they do not change the total amount of kinetic or thermal energy in the gas. Inelastic collisions do exchange the thermal energy of the gas with other forms of energy, such as excitation and ionization energy. Elastic collisions tend to make the gas particles behave as they would in TE, that is, they tend to set up a Maxwellian velocity distribution, while inelastic ones tend generally to destroy such a distribution since they preferentially exchange energies which are at or above certain thresholds. (In strict TE each inelastic collision is exactly balanced by its inverse, so they have no effect.) In most cases of astrophysical interest the elastic collisions are very much more common than inelastic collisions, so it is a good approximation to assume a Maxwellian velocity distribution for the free particles (unless degeneracy effects are important).

The above does not hold if the mean free path for the particles is not small compared with the distance over which the temperature changes appreciably. In this case, particles representing different temperatures are mixed together, and the result will not be the same as in TE at a fixed temperature. It is easy to show, however, that in typical photospheres the particle mean free paths are short enough that particles of significantly different temperatures are very effectively screened from each other. For example, if one assumes the Eddington approximation to the gray temperature distribution, equation (8.11), one finds

$$\frac{1}{T}\frac{dT}{dx} = \frac{k}{4(\tau + 2/3)} . \tag{18.1}$$

k is the volume continuous absorption coefficient and τ is the optical depth. Near the surface, k is of the general order of 10^{-7} cm^{-1} for the solar atmosphere, so a change in depth by some 10^7 cm is necessary in order for the temperature to change by an appreciable fraction of itself. On the other hand, the mean free path for collisions is $1/Na$, where N is the total particle density and a is the collision cross section. With N of the order of 10^{17} ptcls/cm^3 and a of the order of 10^{-16} cm^2, the mean free path is about 10^{-1} cm, which is some 10^8 times smaller than the temperature scale. Thus, collisional transition probabilities are essentially the same as in TE at the same kinetic temperature, and collisions tend to make all $b_j = 1$.

The situation is quite different for radiative transitions. Even if each point in the atmosphere were to emit radiation as in TE at its own local kinetic temperature, the atmosphere is, by definition, composed of those regions where the photon mean free path is *not* small compared to the temperature scale. The photon mean free path is the distance corresponding to optical depth unity, and this is just what equation (18.1) predicts for the temperature scale at the surface. The boundary can be seen from all points in the atmosphere, so atoms are not in an isotropic radiation field. Thus, one might expect radiative transitions in atmospheres to tend to produce populations of the bound levels with $b_j \neq 1$, and the deviations from LTE should become stronger for layers closer to the surface. It is apparent from equation (18.1) that, for very deep regions in a star, the temperature scale becomes much larger than the photon mean free path, the boundary cannot be seen, and radiative as well as collisional transitions tend to produce TE populations of the levels. Thus deviations from LTE are restricted to the outer regions of stars and interstellar space.

The above qualitative arguments will now be supported by a quantitative determination of the population ratio of the bound states and of the line source function. In order to keep the algebra at a reasonable level, an atom with only two energy states will be considered. This analysis was first carried out by R.N. Thomas, *Ap. J.* **125**, 260, 1957. The generalization to a many level atom is straightforward.

Unless the properties of the star in question are changing rapidly with time, the number of atoms in state 1 per unit volume, N_1, does not change with time. Thus the number of transitions $1 \rightarrow 2$ per volume per second must equal the number $2 \rightarrow 1$. As mentioned above, both radiative and collisional transitions occur. The total number of radiative transitions per unit volume and time are found by integrating equations (5.12), but now multiplied by the profile function, over all frequencies and directions. In the case of spontaneous emission, this integration gives N_2A_{21}. In the other two cases, the integration over direction changes I_ν into the mean intensity J_ν. The result is then $N_1B_{12}\overline{J}$ and $N_2B_{21}\overline{J}$ for absorption and induced emission, where $\overline{J}$ is defined by

$$\overline{J} = \int_0^\infty J_\nu \phi_\nu \, d\nu \ . \tag{18.2}$$

Let C_{12} and C_{21} be the probabilities per unit time that an atom will have a collisional transition in the given direction. Then N_1C_{12} is the number of collisional excitations per unit volume and time. Equating the total number of excitations to the number of de-excitations leads to

$$N_1(C_{12} + B_{12}\overline{J}) = N_2(C_{21} + A_{21} + B_{21}\overline{J}) . \tag{18.3}$$

Since the atom is assumed to have only two levels, the only place it can go from level 1 is to level 2, and the only place from which it can enter level 1 is from level 2. For a multi-level atom, one would have to include transitions from level one to all of the other levels, both bound and free.

In TE there is detailed balance, and the rate of collisional excitations and de-excitations exactly balance each other. (If this were not so, equation (18.3) would predict a TE population ratio N_1/N_2 that depends on the relative values of the collisional and the radiative transition probabilities, that is, that depends on the kind of atom that is involved.) Thus $N_1^*C_{12} = N_2^*C_{21}$, where the N^*'s are the TE populations. But it was explained above that the kinetic properties of the gas are the same as if TE were valid, due to the efficiency of the elastic collisions. The C's are then unaffected by deviations from TE. Since the N^*'s satisfy the Boltzmann distribution, one has

$$C_{12} = C_{21} \frac{g_2}{g_1} e^{-u} \ , \tag{18.4}$$

where $u = E_{12}/kT = h\nu/kT$. If the B's are expressed in terms of A_{21} through equations (5.20), and if N_1/N_2 is eliminated through equation (17.6), equation (18.3) leads to the following:

$$\frac{b_1}{b_2} = \frac{C_{21} + A_{21}(1 + c^2\overline{J}/2h\nu^3)}{C_{21} + A_{21}c^2e^u\overline{J}/2h\nu^3} \ . \tag{18.5}$$

Suppose that collisional transitions are very much more frequent than the radiative ones. Then C_{21} is much greater than A_{21}, and equation (18.5) reduces to $b_1 = b_2$, that is, LTE in the line. Now consider the other extreme in which C_{21} is very small compared to A_{21}. Then equation (18.5) yields

$$\frac{b_1}{b_2} = \frac{1}{W}\,(1 - e^{-u} + We^{-u}). \tag{18.6}$$

W is the average mean intensity over the line in terms of the local Planck function:

$$\bar{J} = W\,B_\nu(T) = W\cdot\frac{2h\nu^3/c^2}{e^u - 1}. \tag{18.7}$$

In TE $W = 1$. An absorption line would tend to have W less than unity, although the precise level of the continuum and the kinetic temperature are also involved. Near the surface geometric dilution reduces the value of W because a significant amount of radiation comes only from the lower directions. This geometric effect amounts to a factor of one-half at the surface. One thus expects W to decrease from unity as one approaches the surface from below. Equation (18.6) then predicts (b_1/b_2) to be greater than unity at or just below the surface of the atmosphere: the excited level is less populated than the ground state as compared with TE. This is the same effect as in interstellar space, where the great dilution of the radiation field causes practically all atoms to be in the ground state. The fact that equation (18.6) applies only for the two-level atom means that the more complicated case in which both levels 1 and 2 are excited cannot be predicted from this.

It is a straightforward procedure to substitute equation (18.5) into the expression for the source function (17.7). Define the quantity q as

$$q = \frac{C_{21}(1 - e^{-u})}{A_{21} + C_{21}(1 - e^{-u})}. \tag{18.8}$$

Except for the factor $(1 - e^{-u})$, q is the ratio of the number of collisional de-excitations to the sum of collisional de-excitations and spontaneous emissions. If one defines scattering in the line as photon absorption $1 \rightarrow 2$ followed immediately by spontaneous emission $2 \rightarrow 1$, then q is essentially the probability of non-scattering. The line source function (17.7) then gives

$$S_{12} = qB_\nu(T) + (1 - q)\int_0^\infty J_\nu\phi_\nu\,d\nu. \tag{18.9}$$

Equation (18.9) is analogous to equation (5.24) for the continuum source function. Both expressions have a non-scattering part which is the Planck function of the kinetic temperature, and both have a scattering part which is proportional to the mean intensity. There is a major difference, however, between the two forms of the scattering terms. In arriving at equation (5.24), it was assumed that the scattered photon suffers no frequency change, so the scattering is coherent. While there are a number of processes which will cause small frequency changes, they are of no importance in continuum processes since the latter change only very slowly and smoothly with frequency. For line formation on the other hand, ϕ_ν changes by many orders of magnitude over tiny frequency intervals, and very small frequency shifts between the absorbed and emitted photons cannot be ignored. In equation (18.9) the scattering term is completely non-coherent, which means that the emitted photon has lost all memory of the frequency of the absorbed photon. Notice that in equation (18.9) the source function is essentially independent of frequency over the width of the line.

In the deep layers of the star $J_\nu \to B_\nu$ and $S_{12} \to B_\nu$, regardless of the value of q. As one approaches the surface, J_ν starts to deviate significantly from B_ν and the source function will also become different from the Planck function, the amount depending on the value of q. Eventually one will reach regions where material densities are so low that collisions are not important, and q will there be very small. Thus the scattering term in equation (18.9) must become dominant above a certain level, and there must also be a level above which J_ν is significantly different from B_ν. If the line still has an appreciable amount of absorption above these levels, then LTE does not properly describe the formation of the line.

The kinetic temperature T at a point is determined by the kinetic properties of the particles within a collisional mean free path of that point. As stated earlier, this mean free path is very small on the scale of the atmosphere, so T is determined locally. Likewise, the mean intensity is determined by the properties of the photons within one photon mean free path, so that the whole atmosphere plays an important role in fixing J_ν at any point within the atmosphere.

Another point is important here: while a scattering can change the frequency of a photon by a small amount, the photon must undergo a non-scattering absorption before it can come into equilibrium with its thermal surroundings. If q is very small, the photon has to travel many mean free paths before it thermalizes. In deciding whether radiative processes favor LTE, it is this thermalization length rather than the mean free path which must be compared with the temperature scale height of the atmosphere. LTE is a very great simplification, and because of this there is a tendency to apply it beyond the region in which it is valid.

The two-level atom approximation has been solved in a large number of cases. Notable is a series of papers by J.T. Jefferies and R.N. Thomas in the *Astrophysical Journal* **127**, 667, 1958; **129**, 401, 1959; **131**, 695, 1960. Jefferies and Thomas solve a variety of problems in the Eddington approximation using the method of discrete ordinates for the integral over frequency.

The total source function is the combination of that of the line and that of the continuum. They combine according to equation (5.2). If η is the ratio of line to continuous absorption and if the continuum is formed in LTE, then

$$S_\nu = \frac{1 + \eta q}{1 + \eta} B_\nu + \frac{\eta(1 - q)}{1 + \eta} \int_0^\infty J_\nu \phi_\nu \, d\nu. \qquad (18.10)$$

When the above analysis is generalized to include all levels of an atom, then terms for the collisional and radiative transitions between level 1 and all others must be added. Thus terms which are both non-scattering and non-Planckian appear, and it brings in the ratios b_1/b_j for all j and the $\bar{J}_{1j}$ for all j. The equation describing statistical equilibrium for level 1 is not sufficient to solve the problem, but the equations for the equilibrium of each level must be included, and they must be solved simultaneously. In other words, the formation of a given line cannot be considered by itself; instead, one must consider the simultaneous formation of all possible lines and continua of the given atom. Finally, since a photon cannot be reserved by any element but is liable to be absorbed by any of several, it may be necessary to solve simultaneously for the transitions in more than a single element.

Of course the completely general case mentioned above cannot be solved, and it is not necessary. It is sometimes necessary to carry the approximation well beyond the two-level atom, however. See Jefferies' book *Spectral Line Formation*, especially chapter 8, and Mihalas' book, *Stellar Atmospheres*, especially chapter 13, for further discussion and references.

There is no question that the theory outlined above is basically correct; there is much question, however, about the numerical details of its application. The controversy over the range of validity of LTE has been a long and highly emotional one. A reason for this is that LTE is sometimes applied where it is obviously invalid; another reason appears to be that observations indicate that LTE has such a wide range of validity that it is rather embarrasing to some non-LTE proponents. Except for the centers of very strong lines and lines formed in the chromosphere and corona (the very outer parts of the solar atmosphere), lines observed in the spectra of the Sun and most other stars appear to be quite close to LTE. See the review of B.E.J. Pagel in *Proc. Roy. Soc. A.* **306**, 91, 1968.

The theory outlined in the present section and its application have been one of the main developments in stellar atmospheres since the mid-1950's. While it has not resulted in a major revision of numerical values, it has provided for the first time a solid physical basis for line formation.

19. Line Broadening

The broadening, or profile, function ϕ_ν appears both in the line absorption coefficient and, if LTE is not valid, in the source function. In the present section the principal causes of line broadening are examined so that the resulting forms for ϕ_ν can be derived, at least in rough approximation. There are three main causes of line broadening which are generally important: the Doppler effect, natural broadening, and pressure broadening. These are considered in turn, and then the combined effects are derived.

The Doppler effect causes the frequency of a photon as seen by an observer to differ from that seen by the source. If all speeds are small compared with c, the speed of light, then

$$\Delta\nu = -\frac{\nu v}{c}. \qquad (19.1)$$

v is the relative radial velocity between source and observer, where the sign convention is that v is positive if the two are receeding from each other, so that the observer sees a lower frequency than the source. If the atoms which absorb and emit line radiation have a distribution in radial velocity, then the observer sees a distribution of frequencies absorbed and emitted.

Let $p(v)dv$ be the probability that an atom has radial velocity between v and $v + dv$. There is a one to one correlation between v and $\Delta\nu$ as given by equation (19.1), so

$$\phi_\nu d\nu = -p(v)dv = p\left(-\frac{c\Delta\nu}{\nu}\right)\frac{c\ d\nu}{\nu} . \tag{19.2}$$

If the velocity distribution results from the thermal motions of the atoms, then $p(v)dv$ is given by the one-component Maxwellian distribution:

$$p(v)dv = \frac{1}{\sqrt{\pi}}\, e^{-(v/v_o)^2}\, \frac{dv}{v_o} , \tag{19.3}$$

where

$$v_o = \left(\frac{2kT}{m}\right)^{1/2} \tag{19.4}$$

v_o is the most probable speed of a particle of mass m. It is usual to measure the frequency shift $\Delta\nu$ in terms of the so-called Doppler width D, defined as the Doppler shift due to an atom having a radial velocity v_o:

$$D = \frac{\nu v_o}{c} = \frac{\nu}{c}\left(\frac{2kT}{m}\right)^{1/2} \tag{19.5}$$

The Doppler width D can be measured in wavelength or in frequency units. Which unit is intended should be obvious from the context.

If equations (19.3)-(19.5) are substituted into equation (19.2), one obtains

$$\phi_\nu d\nu = \frac{1}{\sqrt{\pi}}\, e^{-u^2} du , \tag{19.6}$$

where

$$u = \frac{\Delta\nu}{D} \tag{19.7}$$

If, in addition to thermal motions, large scale mass motions exist which cause a dispersion in the radial velocities of the atoms, a model for the form of these motions is necessary to determine the function $p(v)dv$. If the scale of these mass motions is large compared to the length of optical depth unity, the motions are called macroturbulence. If the scale is small the motions are called microturbulence. The reason for this division is that the two extremes affect the radiation in completely different ways. The scale of microturbulence is small compared to the optical depth scale, so it affects the radiation in exactly the same way as do the thermal motions, except that it may have a different velocity distribution. By definition, macroturbulent motions are constant over optical depths of one or two, so the intensity is affected only through a fixed frequency shift. Intensities arising from different regions have different macroturbulent velocities, so this affects the way intensities combine to form a flux: each region will have its own frequency shift, and the flux will be distorted.

It is usual to assume that the microturbulent velocities have a Maxwellian distribution with most probable velocity v_t. Two separate Maxwellian distributions combine to form a new distribution which is also Maxwellian, and the square of the new most probable velocity is the sum of the squares of the old ones. Thus equations (19.6) and (19.7) are still valid, but the Doppler width D is now

$$D = \frac{\nu}{c}\left(\frac{2kT}{m} + v_t^2\right)^{1/2} . \tag{19.8}$$

v_t is generally treated as a new parameter to be determined when calculations are compared with observations. Macroturbulence does not affect the shape of ϕ_ν, but it causes the profile function to be symmetric about a frequency different from ν_o, the normal line center.

The next effect to be considered is known as natural broadening. It is a result of the fact that atoms do not remain indefinitely in their energy states, but the states have finite lifetimes. The uncertainty relation indicates that an energy level is smeared out by an amount that is of the order of $\hbar/T_j = h/2\pi T_j$, where T_j is the time available to measure the energy (the lifetime of the level).

Consider an excited level j of an atom. Let there be $N_j(t_0)$ atoms per unit volume in this level at time t_0. These atoms eventually undergo transitions to other levels. If the atoms are undisturbed, they can only make spontaneous transitions to lower levels. Ignoring transitions to j from higher levels, one has

$$\frac{dN_j}{dt} = -N_j \sum_{i<j} A_{ji} \tag{19.9}$$

The integral of equation (19.9) gives

$$N_j(t) = N_j(t_0)\, e^{-\left(\sum\limits_{i<j} A_{ji}\right) t} \tag{19.10}$$

The average lifetime T_j of the level is

$$T_j = \frac{\int_0^\infty N_j(t)\,t\,dt}{\int_0^\infty N_j(t)\,dt} = \frac{1}{\sum\limits_{i<j} A_{ji}} . \tag{19.11}$$

The energy spread of the level is of the order of

$$\Delta E_j \sim \hbar \sum_{i<j} A_{ji}. \tag{19.12}$$

For the ground state, there are no lower levels. T_1 is essentially infinite, and the level is extremely sharp.

The determination of the shape of an energy level was first made by V. Weisskopf and E. Wigner, *Zs. f. Phys.* **63**, 54, 1930. The energy eigenfunction of state j can be written

$$\psi_j(\mathbf{r},t) = u_j(\mathbf{r})e^{-iE_jt/\hbar} . \tag{19.13}$$

where E_j is the mean energy of level j. The total wave function is

$$\Psi(\mathbf{r},t) = \sum_j a_j(t)\ \psi_j(\mathbf{r},t). \tag{19.14}$$

where $|a_j(t)|^2$ is the probability of finding the atom in state j at time t. Equations (19.10) and (19.11) indicate that spontaneous emission causes the population of a state to decrease with time as e^{-t/T_j}. If the atom has a certain distribution over its states at time t = 0, and only spontaneous transitions are allowed after that time, then for excited states

$$|a_j(t)|^2 = |a_j(0)|^2e^{-t/T_j} , \tag{19.15}$$

$$a_j(t) = a_j(0)e^{-t/2T_j} , \qquad t \geqq 0. \tag{19.16}$$

The wave function for any time $t \geq 0$ is then

$$\Psi(\mathbf{r},t) = a_1(t)\ \psi_1(\mathbf{r},t) + \sum_{j=2} a_j(0)u_j(\mathbf{r})e^{-(iE_j/\hbar + 1/2T_j)t}. \tag{19.17}$$

In the above it is assumed for simplicity that all transitions are back to the ground state j = 1, although this is not an essential part of the derivation. The Fourier transform of the time dependent part of equation (19.17) is

$$Y_j(E) = \frac{1}{\sqrt{2\pi}}\int_0^\infty e^{-(iE_j/\hbar + 1/2T_j)t}\ e^{iEt/\hbar}\ dt , \tag{19.18}$$

or

$$Y_j(E) = \frac{1}{\sqrt{2\pi}}\,\frac{1}{i(E_j - E)/\hbar + 1/2T_j}. \tag{19.19}$$

$Y_j(E)$ must, of course, satisfy

$$e^{-(iE_j/\hbar + 1/2T_j)t} = \frac{1}{\sqrt{2\pi}} \int_{-\infty}^{\infty} Y_j(E)\, e^{-iEt/\hbar}\, \frac{dE}{\hbar} . \tag{19.20}$$

The ψ_j's form a complete and orthogonal set, so from equation (19.17)

$$\int |\Psi|^2 d\mathbf{r} = 1 = |a_1(t)|^2 + \sum_{j=2}^{\infty} |a_j(0)|^2\, e^{-t/T_j} . \tag{19.21}$$

An alternate expression for the same quantity is obtained by substituting equation (19.20) into equation (19.17):

$$\int |\Psi|^2 d\mathbf{r} = |a_1(t)|^2 + \frac{1}{2\pi} \sum_{j=2} |a_j(0)|^2 \int_{-\infty}^{\infty}\int_{-\infty}^{\infty} Y_j^*(E')Y_j(E)\, e^{-i(E-E')t/\hbar}\, \frac{dEdE'}{\hbar^2} . \tag{19.22}$$

Y^* is the complex conjugate of Y. A comparison of the two equations above for each j shows that

$$\int_0^{\infty} |a_j(t)|^2 dt = T_j |a_j(0)|^2 \qquad j \geq 2 \tag{19.23}$$
$$= |a_j(0)|^2 \int_{-\infty}^{\infty} Y_j(E) \frac{dE}{\hbar} \int_{-\infty}^{\infty}\int_0^{\infty} Y_j^*(E')\, e^{-i(E-E')t/\hbar}\, \frac{dE'dt}{2\pi\hbar} .$$

It follows directly from equations (19.18) and (19.20) that the integrals over E' and t in equation (19.23) give simply $Y_j^*(E)$. (If the lower limit for t were $-\infty$ instead of zero, this would be true for an arbitrary function $Y_j^*(E')$.) It is then evident that

$$\frac{1}{T_j} \int_{-\infty}^{\infty} |Y_j(E)|^2 \frac{dE}{\hbar} = 1 . \tag{19.24}$$

The quantity

$$p_j(E)dE \equiv \frac{1}{T_j} |Y_j(E)|^2 \frac{dE}{\hbar} = \frac{1}{2\pi T_j} \frac{dE/\hbar}{(E - E_j)^2/\hbar^2 + 1/4T_j^2} \tag{19.25}$$

is thus the properly normalized probability of finding an atom, known to be in state j, having energy between E and $E + dE$. It has a maximum at E_j, and the total width at half maximum is seen to be $\hbar/T_j$. The finite lifetimes of the states cause them to have a finite spread of energy.

It is usual to define what is called a damping constant:

$$\delta_j = \frac{1}{4\pi T_j} . \tag{19.26}$$

(One also sees the quantity $\gamma = 4\pi\delta$.) Then the energy distribution of level j due to natural broadening is

$$p_j(E)dE = \frac{2\delta_j dE/\hbar}{(E - E_j)^2/\hbar^2 + 4\pi^2\delta_j^2} . \qquad (19.27)$$

Consider now a transition between an initial level i and a final level j. The probability that the involved photon has frequency between ν and $\nu + d\nu$ is equal to the probability that the energy in the final state is $h\nu$ greater (or less) than that from which it started:

$$\phi_\nu \, d\nu = h d\nu \int_{-\infty}^{\infty} p_i(E) p_j(E + h\nu) dE. \qquad (19.28)$$

Substituting equation (19.27) and carrying out the integral, one has

$$\phi_\nu \, d\nu = \frac{\delta}{\pi} \frac{d\nu}{(\nu - \nu_0)^2 + \delta^2} , \qquad (19.29)$$

with

$$\delta = \delta_i + \delta_j . \qquad (19.30)$$

$\nu_0 \equiv (E_i - E_j)/h$ is the frequency of the line center.

When the energy levels are disturbed by nearby particles, the effect is called pressure broadening. To determine this, one must know the type of interaction between the particles, and from this one must be able to find the probability that the energy levels are distorted by a given amount. The frequency shift of an absorbed or emitted photon is $(\Delta E_2 - \Delta E_1)/h$, where the ΔE's are the energy changes of the levels. If r is the distance of a disturbing particle from the emitting or absorbing atom, the interaction is usually written in the form

$$\Delta\nu = \frac{C}{r^n} . \qquad (19.31)$$

The constant C depends on the levels and on the type of interaction. An approximate description of the different types of interaction follows.

Let the atom of interest be represented as a dipole of moment $\mathbf{p}$. If an external electric field $\mathbf{F}$ is applied, the dipole moment becomes

$$\mathbf{p} = \mathbf{p}_0 + \alpha\mathbf{F}. \qquad (19.32)$$

$\mathbf{p}_0$ is the intrinsic moment of the atom, and the second term in (19.32) is the moment induced by the field, where α is the polarizability of the atom. The energy change is proportional to

$$\Delta E = h\Delta\nu \sim \int \mathbf{p} \cdot d\mathbf{F} = \mathbf{p}_0 \cdot \mathbf{F} + \frac{1}{2} \alpha F^2 . \qquad (19.33)$$

When the disturbing particle is charged, that is, a free electron or an ion, the field is proportional to r^{-2}, and the interaction is known as the Stark effect. If the perturbing particle is neutral, its main effect is through its dipole field which falls off with distance as r^{-3}.

Only hydrogen-like atoms have an appreciable intrinsic dipole moment, so for them the first term in equation (19.33) is dominant (unless the field is extremely large); for other atoms, the second term is generally dominant. The electric fields of charged particles are much greater than those of dipoles, so the latter can usually be neglected if there is appreciable ionization. A neutral particle needs to have a dipole (or higher order) moment in order to disturb other atoms. Since hydrogen is usually the most abundant neutral particle at the lower temperatures, it is generally the only neutral particle of importance in astrophysics in broadening the lines of other atoms.

The above information can be combined with equations (19.31) and (19.33) to yield the following table of results:

Absorber	Disturber	n	Name of Effect
Hydrogen-like	Neutral	3	Self or resonance broadening
Hydrogen-like	Charged	2	Linear Stark effect
Not hydrogen-like	Neutral	6	Van der Waals broadening
Not hydrogen-like	Charged	4	Quadratic Stark effect

The case for $n = 3$ is generally known as self broadening, since this holds for any neutral particle disturbing an atom of the same kind. As stated above, this is important in astrophysics only for hydrogen. Note that all atoms become hydrogen-like for large enough values of the principal quantum number of the outer bound electron, so the Stark effect changes from quadratic to linear as one considers the higher levels of any atom.

When the important interactions for a given line have been identified, their net effects on the line must be determined. This has not been carried out under the most general conditions, but only in the limits of two extreme cases. When the collision time t_c is very short compared with the average radiation time of a line t_r, the time dependent behavior of the emitting atom during a collision can be ignored. Only the collision as a whole need be considered. This approximation is variously known as the collision, the impact, or the phase shift theory. When the radiation time is very short compared with the collision time, the time dependent behavior of the perturbers during the absorption or emission can be ignored. One need only calculate the probability that emitting atom finds itself in a static field of given size. This approach is known as the statistical or static field approximation. Thus the theory of pressure broadening has been highly developed only for $t_c \gg t_r$ and for $t_r \gg t_c$.

The collision time t_c is of the order of d/v, where d is the impact parameter (closest distance of approach) of the collision, and v is the relative velocity. The average radiation time of a line t_r is defined as $1/2\pi\Delta\nu_0$, where $\Delta\nu_0$ is the mean frequency shift from the line center. The ratio then is

$$\frac{t_r}{t_c} \simeq \frac{v}{2\pi\Delta\nu_0 d} . \tag{19.34}$$

High temperatures and small masses, both of which increase v, tend to favor the collisional approximation, while very broad lines with large $\Delta\nu_0$ tend to favor the statistical theory. One can go beyond the above and actually define a radiation time for each photon absorbed or emitted. Using $t_r = 1/2\pi\Delta\nu$, where now $\Delta\nu$ is the shift of a given photon, one can evaluate t_r/t_c for each photon. One finds a condition on $\Delta\nu$ for the limiting case where $t_r = t_c$. If equation (19.31) is used to eliminate d, the limiting frequency shift is found to be

$$\Delta\nu_1 \approx \left(\frac{v}{2\pi C^{1/n}}\right)^{n/(n-1)} \tag{19.35}$$

For shifts much greater than $\Delta\nu_1$ the static theory is applicable; for those much smaller than this, the impact theory is valid. It appears that free electrons have such large mean velocities in stellar atmospheres that electron broadening is usually collisional, but note the discussion of hydrogen lines at the end of this section. The more massive ions have lower thermal velocities, and the situation is not clear-cut. The quadratic Stark effect seems to be somewhat borderline, with the linear Stark effect (n = 2) favoring the statistical theory and the van der Waals broadening (n = 6) tending toward the collisional theory. Unsöld in his book *Physik der Sternatmosphären* gives numerical values of the limiting shift for some examples of the linear Stark effect (on page 322) and of the quadratic Stark effect (on page 327). Unsöld uses a slightly different constant from the one appearing in equation (19.35).

For purposes of illustration simple examples of the limiting theories will now be given. Much more complete discussions are given, for example, in H.R. Griem, *Plasma Spectroscopy*, McGraw-Hill, 1964; C.R. Cowley, *The Theory of Stellar Spectra*, Gordon and Breach, 1970; and G. Traving, *Druckverbreiterung von Spektrallinien*, Braun, 1960.

The simplest example of a statistical theory is known as the nearest neighbor approximation. It is assumed that only the nearest disturbing particle need be taken into account. If N is the number of disturbers per unit volume, then $4\pi Nr^2dr$ is the probability of finding such a particle between r and r + dr. Let $p^*(r)$ be the probability that no particle is within r; then

$$p^*(r + dr) = p^*(r)(1 - 4\pi Nr^2dr).$$

This can be integrated to yield

$$p^*(r) = e^{-(r/r_0)^3}, \tag{19.36}$$

where r_0 is the mean distance between particles:

$$r_0 = (\tfrac{4}{3}\pi N)^{-1/3}. \tag{19.37}$$

The probability $p(r)dr$ of finding the nearest particle between r and r + dr is

$$p(r)dr = p^*(r)4\pi Nr^2dr = 3\,\frac{r^2}{r_0^2}\,e^{-(r/r_0)^3}\,\frac{dr}{r_0}. \tag{19.38}$$

A disturbing particle at r produces a frequency shift given by equation (19.31). If r is eliminated from equation (19.38), one obtains the broadening function:

$$\phi_y dy = \frac{3}{n} e^{-(y^{-3/n})} y^{-(n+3)/n} dy, \tag{19.39}$$

where

$$y = \frac{\Delta\nu}{\Delta\nu_0}, \tag{19.40}$$

and

$$\Delta\nu_0 = \frac{C}{r_0^n} \tag{19.41}$$

Note that y is always positive. This means that all shifts are in the same direction, either to larger or to smaller frequencies, depending on the sign of the interaction constant C. The broadening is thus not symmetric about the line center. If the energy levels are degenerate, each separate line component has its own value of C, and the net line broadening must be calculated component by component.

The nearest neighbor approximation is valid only for very close encounters, so equation (19.39) holds in practice only for large values of y. In this region it is seen that ϕ_ν is proportional to $(\Delta\nu)^{-(n+3)/n}$. If all of the disturbing atoms are taken into account instead of only the nearest, the generalization of the above analysis leads to what is known as the Holtsmark distribution. For the latter see the above-mentioned references or S. Chandrasekhar, *Rev. Mod. Phys.* **15**, 1, 1943.

Perhaps the simplest example of a collisional type theory is one in which the atom is pictured as indefinitely radiating a wave of fixed frequency ν_0, but in which the wave is intermittantly interrupted by collisions. This satisfies the requirement of the collision approximation that the radiation time be much longer than the collision time.

If each collision completely disrupts the wave train, the train is a series of independent waves of duration equal to the free time between collisions. Let t' be the time between collisions; then the wave is of the form $e^{-i\omega_0 t}$ for times in the interval $0 \le t \le t'$. The Fourier transform of the wave is

$$F(\omega,t') = \frac{e^{i(\omega-\omega_0)t'} - 1}{\sqrt{2\pi}\, i(\omega-\omega_0)}. \tag{19.42}$$

The energy carried by waves of frequency ω is proportional to the square of the transform $|F(\omega,t')|^2$. The net energy carried by waves of this frequency is obtained by multiplying the above by the probability that the atom experiences a free time between collisions of t' and integrating the result over all t'. Equation (2.4) gives the probability that a photon will travel the distance between s and s + ds before being absorbed, where k is the reciprocal of the mean free path. If one replaces the distance s by the time interval t' between collisions, the analysis leading to equation (2.4) also determines the probability of an atom having a free time between t' and t' + dt'. The result is identical to equation (2.4) except that k is replaced by the reciprocal of the mean free time T_c between collisions:

$$p(t')dt' = e^{-t'/T_c} \frac{dt'}{T_c} . \tag{19.43}$$

One finds

$$|F(\omega)|^2 = \int_0^\infty p(t')|F(\omega,t')|^2 \, dt' = \frac{1}{2\pi} \frac{1}{(\omega - \omega_0)^2 + 1/T_c^2} . \tag{19.44}$$

The broadening function is found by normalizing the above. Introducing the collisional damping constant

$$\delta_c = \frac{1}{2\pi T_c} , \tag{19.45}$$

one finds the following for the broadening function:

$$\phi_\nu \, d\nu = \frac{\delta_c}{\pi} \frac{d\nu}{(\nu - \nu_0)^2 + \delta_c^2} . \tag{19.46}$$

The above model for pressure broadening was considered by H.A. Lorentz in the early part of the present century. The function (19.46), which is of exactly the same form as equation (19.29) for natural broadening, is often known as the Lorentz profile. A more general derivation by E. Lindholm indicated that collisions cause a shift in the position of the line center as well as a broadening of the line. More recent quantum mechanical calculations indicate that non-adiabatic effects, collisionally induced transitions in the radiating atom, can be quite important under some circumstances.

A given line is broadened by several effects simultaneously, and the separate effects must be combined. A common example is one in which the pressure broadening is of the Lorentz type. In the derivation of equation (19.29) it was seen that two Lorentz profiles combined to produce a new Lorentz profile in which the new damping constant is the sum of the old ones. In this case equation (19.29) represents the combined effects of both pressure and natural broadening, with δ being the sum of the natural and collisional damping constants. Of course, the collisional damping constant may itself be the sum of those of separate collision mechanisms, as long as each of them can be adequately represented by a Lorentz profile.

The final profile function is obtained by folding together the Lorentz function and the Doppler function (19.6). Consider an atom with the instantaneous radial velocity v. It absorbs and emits according to the Lorentz profile, but the frequency observed will be Doppler shifted according to equation (19.1):

$$\phi_\nu(v)d\nu = \frac{\delta}{\pi} \frac{d\nu}{(\nu - \nu_0 v/c - \nu_0)^2 + \delta^2} . \tag{19.47}$$

Multiply this by the probability of v , equation (19.3), and integrate over all radial velocities:

$$\phi_\nu d\nu = \phi_u du = \frac{a du}{\pi^{3/2}} \int_{-\infty}^{\infty} \frac{e^{-y^2} dy}{(u + y)^2 + a^2} . \tag{19.48}$$

The substitution $y = v/v_0$ is used. u is the frequency shift in units of the Doppler width D as in equation (19.7), and a is the damping constant, in the same units of the Doppler width:

$$a = \frac{\delta}{D} . \tag{19.49}$$

The form of equation (19.48) is known as the Voigt profile. It is common to see the function $H(a,u) \equiv \sqrt{\pi}\, \phi_u$ in the literature. The H function is not normalized to unity for integration over u, but it does equal unity at the line center in the limit of very small damping. The Voigt profile is tabulated in, for example, G.D. Finn and D. Mugglestone, *M. N.* **129**, 221, 1965 and D.G. Hummer, *Mem. R.A.S.* **70**, 1, 1965.

Hydrogen line broadening is much more complicated than the above. On the basis of the previous discussion, one would expect statistical broadening by ions and collisional broadening by electrons. At lower temperatures self broadening might be important as well. The electron effect is complicated by non-adiabatic effects, and the pressure broadening profile is not of the Lorentz shape.

Edmonds, Schlüter, and Wells, *Mem. R.A.S.* **71**, 271, 1967, suggest that the static field approximation is valid for electrons as well as ions in determining hydrogen line profiles. The numerical coefficient in equation (19.35) for the boundary between the static field and the collision regimes is not well determined. For the stronger lines, the transition to the static field case may occur while there is still appreciable line absorption.

As will be shown in the next section, the central regions of a line (within about three Doppler widths of the center) are dominated by Doppler broadening. Beyond this region, natural and pressure broadening are dominant. For small frequency shifts and for weak lines, therefore, it is not necessary to know the pressure broadening accurately, and the inadequacies in the theory are not important.

20. Profiles, Equivalent Widths, Curves of Growth

The various quantities needed to solve the line problem have been discussed in previous sections. It is now a matter of putting the pieces together. The intensity emitted normally by a plane atmosphere is

$$I_\nu = \int_0^\infty S_\nu(\tau) e^{-\tau} d\tau . \tag{20.1}$$

One could write the equation for other geometries and for other directions of emergence, but these are unnecessary complications in the present context. Equation (20.1) is valid for both line and continuum radiation.

One generally starts the line problem with what one hopes is an accurate model atmosphere. By the methods outlined in Chapter 3, one has obtained a table of physical conditions as functions of τ_c, the optical depth in the continuum in the neighborhood of the line being analysed. If the line is assumed to be formed in LTE, the line source function is immediately known at each point in the atmosphere; however, one needs to find the line absorption at each point before equation (20.1) can be solved for the line radiation.

The atomic absorption coefficient for the line is given by equation (17.4), so it is necessary to calculate the broadening function ϕ_ν for each point in the atmosphere. The correction for induced emission must also be found if it is important. For an assumed abundance of the element in question, the contribution of the line absorption k_l to the total volume absorption coefficient k at any frequency in the neighborhood of the line is calculated to obtain $k = k_l + k_c$. If the total volume absorption coefficient is determined, one can integrate through the atmosphere to find τ, the total optical depth. The physical conditions are now known as functions of τ, and equation (20.1) can be integrated to yield the theoretical line intensity. When this process has been repeated for a number of frequencies across the line, a complete picture of the line shape or profile emerges.

If LTE is not a satisfactory approximation for the line, the solution becomes much more difficult. The source function is given by equation (18.10), or a suitable modification if the two level relation is insufficient. The source function depends on the radiation field which is being sought, so an iterative scheme might be necessary. If the lower level of the line has its population strongly influenced by transitions to other levels, the simultaneous solution for these several transitions must be made. The absorption coefficient as well as the source function is influenced by non-LTE effects, as it depends on the excitation and ionization conditions.

The calculated line profile depends upon many things: the model atmosphere, the broadening function, the mechanism of line formation, the assumed excitation and ionization conditions, and the abundance of the element. It is also necessary for the various atomic constants of the line to be known, such as the oscillator strength, damping constants, and possibly other special radiative and collisional transition probabilities. A comparison of observed and calculated line profiles provides a very complicated test of the accuracy with which the assumed values of parameters and the assumed theories of the physics represent the actual conditions in the star. The observational check can be made more meaningful by using many lines and, in the case of the Sun, by using the same lines at different positions on the disk. Of course the quality of the observations is also an important part of the test.

For many years it was not possible to obtain very accurate observations of line profiles, and most observational checks were made with the total strength of a line rather than with the detailed shape. While this observational limitation no longer exists, it is still much easier to measure the line strength than its profile. Further, the line strength is much less sensitive to some of the uncertainties which enter the line formation theory than is the profile; if these uncertainties are not the direct objects of interest in an investigation, therefore, it would be advantageous to use only the total strengths of the lines.

The strength of a line is measured by its equivalent width W, abbreviated ew:

$$W = \int_0^\infty \frac{I_c - I_\nu}{I_c} d\nu. \tag{20.2}$$

I_c is the intensity in the continuum in the neighborhood of the line, and I_ν is the intensity in the line itself; both are monochromatic quantities. The ew W is measured in frequency units (wavelength units can also be used), but it is not a monochromatic quantity. W can be measured in terms of the intensity, as above, or in terms of the flux. Note that W is not a measure of the absolute strength of a line, but of its strength relative to that of the background continuum. The integrand of equation (20.2) is known as the residual intensity r:

$$r = \frac{I_c - I_\nu}{I_c} \tag{20.3}$$

To illustrate how ew's are less sensitive to certain quantities than are profiles, it will now be shown than an ew does not depend on macroturbulent velocities in the source. As indicated in the last section, macroturbulence is the term given to motions which have a scale much larger than the distance corresponding to optical depth unity. It is a property of fluxes, not intensities, so that flux ew's are used in this calculation.

The emitting surface of the source is divided into n separate regions indicated by the superscript i, $i = 1,2,\ldots,n$. Each region is to be large enough to be optically thick, yet small enough to have an essentially constant turbulent velocity. The definition of macroturbulence makes this construction possible. The regions are optically thick, so they act as independent sources, and the total flux is the sum of the individual fluxes:

$$F_c = \sum_{i=1}^{n} F_c^i, \qquad F_\nu = \sum_{i=1}^{n} F_\nu^i. \tag{20.4}$$

The residual flux in the total radiation is

$$r = 1 - \frac{F_\nu}{F_c} = 1 - \frac{\sum_i F_\nu^i}{F_c}, \tag{20.5}$$

In terms of quantities for each individual region, $F_\nu^i = F_c^i - F_c^i r^i$. If this is substituted into equation (20.5), the result is

$$r = \frac{\sum_i F_c^i r^i}{F_c}. \tag{20.6}$$

As far as the line is concerned, F_c and other continuum quantities do not depend upon frequency; therefore, the total ew is

$$W = \frac{1}{F_c}\sum_i F_c^i \int_0^\infty r^i \, d\nu = \frac{1}{F_c}\sum_i F_c^i W^i \tag{20.7}$$

The ew of the total flux is the average of the individual ew's, weighted by the individual fluxes. The W^i's are independent of the macroturbulent velocities, and W must be also. The line profile is distorted by this velocity field, but the ew is not affected. Stellar rotation produces a velocity dispersion among the emitting surface elements that has the same properties as the macroturbulence assumed above; therefore rotation of the stars makes lines broader and more shallow, but it does not change the ew's of the lines. Microturbulence is quite different. Regions small enough to have a uniform microturbulent velocity are optically thin; they partially shield each other, and their fluxes are not simply additive.

It is not worthwhile to examine here in detail the numerical techniques that are used to calculate line profiles and ew's. It is only useful to summarize some of the basic properties of lines in stellar atmospheres. An oversimplified model of line formation will be used for illustration in order to keep the main physical principles from being hidden by the mathematical details. The so-called Milne-Eddington model for line formation and LTE are assumed.

The first assumption of the Milne-Eddington approximation is that the continuum source function is a linear function of the continuum optical depth:

$$B_\nu(T) = B_0(b\tau_c + 1). \tag{20.8}$$

Since LTE is assumed, $B_\nu(T)$ is the source function in both continuum and line. While there is no strong theoretical justification for the form of equation (20.8), it is an interpolation formula which contains one of the essential properties of a stellar atmosphere, namely, a temperature increase with depth. Its simple form makes it easy to use from a mathematical viewpoint. B_0 and b are adjustable constants which can be used to improve the fit. Features which depend on the reversal of the temperature gradient in the outer atmosphere obviously cannot be reproduced by the present model.

The second main assumption of the Milne-Eddington approximation is that the ratio of line to continuum absorption does not depend on depth:

$$\eta = \frac{k_l}{k_c} \neq f(\tau). \tag{20.9}$$

It follows from equation (20.9) that $\tau_l = \eta\tau_c$, and the total optical depth is

$$\tau = \tau_l + \tau_c = (1 + \eta)\tau_c. \tag{20.10}$$

Equation (20.8) is a reasonably good approximation, at least over a limited range of depths. Equation (20.9) is quite good for some lines, very poor for others. Both relations are artificial and are made only for mathematical convenience: they allow the line problem to be easily solved without doing very much violence to the physical principles involved. In fact very useful quantitative data can be obtained by comparing real stars with Milne-Eddington models; this is an advantage that one does not expect a priori.

If equations (20.8) and (20.10) are substituted into equation (20.1), one finds

$$I_\nu = B_0\left(\frac{b}{1+\eta} + 1\right)$$

$$I_c = B_0(b + 1). \tag{20.11}$$

The residual intensity is

$$r = \frac{b}{1+b}\,\frac{\eta}{1+\eta}\,. \tag{20.12}$$

The line absorption affects the radiation only through η, that is, through the ratio of line to continuous absorption. For very weak lines, $\eta \to 0$ and $I_\nu \to I_c$. For very strong lines, $\eta \to \infty$ and $I_\nu \to B_0$, the surface value of the source function. In the strong line case the residual intensity r depends only on the constant b, which fixes the rate of increase of the source function with depth. Note that for an isothermal atmosphere, $b = 0$ and $I_\nu = I_c$. If the source function is everywhere the same, it makes no difference whether the radiation originates in deep or in shallow layers.

In order to be more specific about the line properties, the shape of η with frequency needs to be known. It is usual to set

$$\eta = \eta_0\, f(u), \tag{20.13}$$

where, as usual, u is the distance from the line center measured in terms of the Doppler width, and η_0 is the value of η at the line center. Thus $f(0) = 1$, and f is proportional to the broadening function ϕ_u. For a line having a Voigt profile, $f(u) = H(a,u)$ if the damping constant a is very small, where H is $\sqrt{\pi}$ times the quantity given by equation (19.48).

The Voigt profile appears to be appropriate for many lines, and it will also be assumed here; however, in keeping with the other simplifications made to hold the physics in full view and to put the mathematics in the background, an approximate form is used.

In most astrophysical applications, the damping constant a in equation (19.48) is very small, usually in the range 10^{-3} to 10^{-1}. Consider equation (19.48) first for moderately small values of u. The denominator of the integrand has a sharp minimum at $y = -u$, and most of the contribution to the integral comes from y values very near $-u$. (If a were not very small, a much wider range of y values would be important.) Thus e^{-y^2} is nearly constant over the range of importance, and ϕ_u is approximately given by

$$\phi_u \simeq \frac{ae^{-u^2}}{\pi^{3/2}} \int_{-\infty}^{\infty} \frac{dy}{(u + y)^2 + a^2} = \frac{1}{\sqrt{\pi}} e^{-u^2} . \qquad (20.14)$$

This is exactly the form of the profile function for thermal Doppler broadening alone, as indicated by equation (19.6); therefore, near the line center Doppler broadening dominates over the effects of pressure and natural broadening.

Consider now equation (19.48) for very large values of u, that is, far from the line center. The denominator still has a sharp minimum at $y = -u$, but the factor e^{-y^2} is now so small that that region can no longer be of importance to the integral. Instead the main contribution comes from much smaller values of y (in absolute value), so the denominator is approximately equal to u^2. Then

$$\phi_u \simeq \frac{a}{\pi^{3/2} u^2} \int_{-\infty}^{\infty} e^{-y^2}\, dy = \frac{a}{\pi u^2} . \qquad (20.15)$$

A comparison of this expression with equation (19.29) indicates that this is what ϕ_u reduces to for large u for a pure Lorentz shape. It follows that for large distances from the line center, natural and pressure broadening dominate over the Doppler effect.

The transition region between the extremes of equations (20.14) and (20.15) can be roughly found by equating the two expressions. One finds that the transition occurs near $u = 3$ for the damping constant a in the range of 10^{-3} - 10^{-1}, and it is very insensitive to the precise value of a. (The form of equation (20.15) dominates over equation (20.14) for both large and small values of u; however, equation (20.15) is valid only for large u.) The part of the line with $u \lesssim 3$ is known as the Doppler core of the line; beyond are the wings of the line. Only very strong lines have significant wings: the line absorption at $u = 3$ is about $e^{-9} = 10^{-4}$ times that at the line center, so the line must be extremely strong if there is to be appreciable absorption beyond $u = 3$.

The expression (20.13) can then be approximately written as follows:

$$\eta \simeq \eta_0 e^{-u^2} , \qquad u \lesssim 3$$

and

$$\eta \simeq \frac{a\eta_0}{\sqrt{\pi} u^2} , \qquad u \gtrsim 3 . \qquad (20.16)$$

The approximate line profiles are found by substituting into equation (20.11).

It is convenient for purposes of illustration to divide lines into three classes: weak, medium, and strong. Weak lines are those for which the absorption is small even in the line center, that is, $\eta_0 \ll 1$. Medium lines have strong absorption at the center, but they become weak within the Doppler core. The absorption at the edge of the core, $u \simeq 3$, is about 10^{-4} of that at the center, so for medium lines $1 \ll \eta_0 \lesssim 10^4$. Strong lines are those with well developed wings, $\eta_0 \gtrsim 10^4$.

Equation (20.12) yields the following for the residual intensity at the center of the line:

$$r_0 = \frac{b}{1 + b}\eta_0 \qquad \text{(weak)}$$

$$r_0 = \frac{b}{1 + b} \quad . \qquad \text{(medium and strong)} \qquad (20.17)$$

The central depth is proportional to the central absorption η_0 for weak lines, but it is independent of η_0 for medium and strong lines. This result is expected: when η_0 has reached a certain large value, the radiation from the line center comes essentially from the surface of the star. Increasing η_0 further cannot affect the central radiation, as the emitted radiation will still come from the surface. If, instead of equation (20.8), one had a temperature increase in the outer atmosphere, simulating the chromosphere of the Sun, the result would be quite different. Weak lines would not "see" the temperature rise because it would be transparent to their radiation; however, the centers of very strong lines would see such regions. If the lines were formed in LTE, such lines would show emission peaks in their cores. If non-LTE effects are dominant, the source function is not particularly sensitive to the increase in the kinetic temperature, and an emission core may not occur even in lines which are strong enough to see the region of the temperature rise. The understanding of which lines should show these emission cores and which ones should not was one of the early triumphs of the non-LTE work (see R.N. Thomas, *Ap.J.* **125**, 260, 1957).

The above illustrates a more general point. From equation (20.1), if the source function is constant with depth, $I_\nu = S_\nu$. If S_ν does vary with depth, the emergent intensity is an average of the source function over the depths of importance to the radiation escaping from the star. Since one mean free path below the surface is an average depth of formation of the emitted radiation, one expects the emergent intensity to be rather close to the source function at that depth. This is a very good rule of thumb which gives useful results when only rough calculations are sufficient.

As indicated in Section 8, the above rule of thumb is exact if the source function is a linear function of optical depth. The Milne-Eddington source function (20.8) is linear, and the intensities in equations (20.11) are seen to be equal to the source function at a depth where the appropriate optical depth is unity. The residual intensities in equations (20.17) can also be understood from the same argument.

A measure of the width of a line is given by u^*, the value of u for which the residual intensity falls to one-half its central value. One has from equations (20.12) and (20.13)

$$\eta(u^*) = \frac{\eta_0}{2 + \eta_0} \ . \qquad (20.18)$$

For the three classes of lines,

$$u^* \approx 1 \qquad \text{(weak)},$$

$$u^* \approx (\ln \eta_0)^{1/2} \qquad \text{(medium)}, \qquad (20.19)$$

and

$$u^* \approx (a\eta_0)^{1/2} \qquad \text{(strong)}.$$

The line widths behave nearly the opposite of the residual intensities given in equation (20.17): weak line widths are independent of η_0, medium line widths are very insensitive to η_0, and strong lines have widths which are proportional to the square root of the central absorption and of the damping constant.

Weak and medium lines are broadened by the Doppler effect alone. Only an extremely small per cent of the atoms have radial velocities greater than two or three times the average, which means that only this small per cent can absorb at frequencies having u more than two or three. There are so few particles having extreme radial velocities that increasing η_0 does not significantly increase the absorption beyond about $u = 3$, and the line width is essentially independent of the central absorption η_0. When damping becomes more important than the Doppler effect, however, an atom does not need a large velocity to absorb at large u, and wide wings grow on the line as the strength increases.

Equations (20.17) and (20.19) show how the shapes of the lines vary with central absorption. In the weak line stage, the depth increases with η_0 but the width remains essentially constant. Eventually the central depth approaches its saturation value given by the second relation in equation (20.17), and the line width very slowly starts to increase. Finally the strong line regime is reached. The central depth remains fixed, but the development of the damping wings causes the line to widen quite appreciably.

The ew is

$$W = \int_0^\infty r d\nu = D\int_{-\infty}^{\infty} r du = \frac{Db}{1+b}\int_{-\infty}^{\infty} \frac{\eta \, du}{1+\eta} . \qquad (20.20)$$

For weak lines the η in the denominator can be neglected, and equation (20.16) yields

$$W = \frac{\sqrt{\pi} Db}{1+b}\eta_0 \qquad \text{(weak)}. \qquad (20.21)$$

Let u_1 be the value of u for medium and strong lines for which $\eta = 1$. As u gets larger than u_1, the line absorption rapidly goes to zero and can, in a rough approximation, be neglected. For essentially all u less than u_1, η is very large; therefore, equation (20.20) gives

$$W \simeq \frac{Db}{1+b}\int_{-u_1}^{u_1} du = \frac{2Dbu_1}{1+b} \qquad \text{(medium and strong)}. \qquad (20.22)$$

Using the appropriate form of equation (20.16), one finds that u is essentially the same as u*:

$$u_1 = (\ln \eta_o)^{1/2} \qquad \text{(medium)}$$
$$u_1 = \left(\frac{a\eta_o}{\sqrt{\pi}}\right)^{1/2} \qquad \text{(strong)}. \qquad (20.23)$$

The ew's of these two classes of lines are then approximately

$$W = \frac{2Db}{1+b}(\ln \eta_o)^{1/2} \qquad \text{(medium)}$$
$$W = \frac{2Db}{1+b}\left(\frac{a\eta_o}{\sqrt{\pi}}\right)^{1/2} \qquad \text{(strong)}. \qquad (20.24)$$

In equations (20.21) and (20.24) W and D are measured in the same units, wavelength or frequency.

The relation between ew and central absorption η_o is known as the curve of growth. It indicates how the absorbed energy grows as the number of absorbing particles increases. For weak lines the ew is proportional to the central absorption. In the medium region, the line core saturates and the ew becomes essentially independent of the central absorption. For strong lines the damping wings develop and ew grows as the square root of the central absorption. The numerical coefficients appearing in the above equations are not accurately determined; however, the general relations between the physical variables have a validity which is much broader than would be indicated by the idealized model for which they were derived.

The curve of growth gives information on the change of the ew of a line when the number of absorbing atoms changes. Its usefulness is greatly increased by the fact that, if certain quantities are suitably standardized, different lines can be plotted on the same curve of growth. This makes it possible to obtain a form of the curve of growth directly from observations. By fitting this curve onto a theoretical curve, important data concerning the star can be determined.

From the definitions of the quantities that go into η_o, noting in particular equation (17.4), one can write

$$\log \eta_o = \log \frac{N}{N_H} - \log\left(\frac{D(\nu)}{\nu_o}\right) + \log z , \qquad (20.25)$$

where

$$\log z = -1.82 + \log\left(\frac{h(T,P_e)}{k_H B_I}\right) + \log(1 - e^{-\chi}) + \log\left(\frac{gf}{\nu_o}\right) - \frac{5040\chi}{T} . \qquad (20.26)$$

N and N_H are the total abundances of the element in question and of hydrogen, respectively, $h(T,P_e)$ is the fraction of the atoms of the element which are in the ionization stage giving rise to the line in question, k_H is the volume continuous absorption coefficient in the neighborhood of the line per hydrogen atom, B_I is the partition function of the element, $x = h\nu_o/kT$, g is the statistical weight of the lower level, f is the f-value of the line, and χ is the excitation potential of the lower level. χ is measured in electron volts, but other quantities are in cgs units.

If F and G are defined by

$$G \equiv \int_{-\infty}^{\infty} \frac{\eta}{1+\eta}\, du \tag{20.27}$$

$$F \equiv \left(\frac{1+b}{b}\right) \frac{W(\nu)}{\nu_o}, \tag{20.28}$$

one sees from equation (20.20) that

$$\log F = \log G + \log\left(\frac{D(\nu)}{\nu_o}\right). \tag{20.29}$$

The relation between $\log G$ and $\log \eta_o$ is what can be called the theoretical curve of growth. It depends only upon the broadening function ϕ_ν; for a given type of broadening, for example, the Voigt profile, the theoretical curve can be determined once and for all.

For any line which has a measured ew, F can be found. (The constant b needs to be known also; a crude model atmosphere of the star suffices to find b for any wavelength region, and models are available for essentially all types of stars.) z can be found from the atomic constants of the line. (Rough values of T and P_e, average temperature and electron pressure in region of formation, are needed; again, a rough model atmosphere yields these.) A plot of $\log F$ vs $\log z$ for a number of lines produces the observational curve of growth.

Consider the quantity

$$\frac{D(\nu)}{\nu_o} = \frac{1}{c}\left(\frac{2kT}{m} + v_t^2\right)^{1/2}, \tag{20.30}$$

(see equation (19.8).) For lines of a given element, the above varies only through T and the microturbulent velocity v_t. The variation is very small through the line forming region, however, and to a good approximation $D(\nu)/\nu_o$ is a constant. Thus the differences $\log \eta_o - \log z$ and $\log F - \log G$ are nearly constants; this in turn means that the theoretical and the observational curves have the same shape: fitting them together then determines the values of the Doppler width (and, therefore, the microturbulent velocity v_t) and the abundance ratio (N/N_H).

The analysis described above is very rough; however, it is very easily carried out and it produces results of surprisingly high accuracy. Many of the numerical details are found in *Ann. d'Ap.* **30**, 659, 1967.

There are many forms of analysis under the name of curve of growth techniques. Not all of them share the over-simplified assumptions which have been used in the present section. The analysis given above represents, in my opinion, about the most painless way a novice in this field can learn the basic principles of line formation. The person who wishes to become scientifically active in this field must, of course, go into many of the details which have been omitted here.

APPENDIX 1

PHYSICAL AND ASTRONOMICAL CONSTANTS

Speed of light	c =	2.9979×10^{10} cm sec^{-1}
Boltzmann's constant	k =	1.3805×10^{-16} erg °K^{-1}
Planck's constant	h =	6.6256×10^{-27} erg sec
Electron charge	e =	4.8030×10^{-10} g$^{1/2}$cm$^{3/2}$sec^{-1}
Electron mass	m_e =	9.1091×10^{-28} g
Proton mass	m_p =	1.6725×10^{-24} g
Mass of unit atomic weight	m_u =	1.6604×10^{-24} g
Gas constant	R =	8.3143×10^{7} erg °K^{-1} mol^{-1}
Gravitational constant	G =	6.670×10^{-8} g^{-1}cm^{3}sec^{-2}
Stefan-Boltzmann constant	σ =	5.6697×10^{-5} erg cm^{-2} sec^{-1}°K^{-4}
Bohr radius	a_0 =	5.2917×10^{-9} cm
Avogadro's number	N_A =	6.0225×10^{23} atoms mol^{-1}
Mass of Sun	M =	1.989×10^{33} g
Radius of Sun	R =	6.960×10^{10} cm
Luminosity of Sun	L =	3.90×10^{33} erg sec^{-1}
Effective temperature of Sun	T_e =	5800 °K
Surface gravity of Sun	g =	2.739×10^{4} cm sec^{-2}

1 eV = 1.60210×10^{-12} erg
= energy of photon of wavelength 1.23981×10^{-4} cm
= energy of photon of frequency 2.41804×10^{14} sec^{-1}

APPENDIX 2

PROBLEMS

1. A sphere is a black body at temperature T_1. It has radius R and is on the y-axis distant r_1 from the origin ($r_1 >> R$). A cube is also a black body, having temperature T_2. Its edges are of length s and are parallel to the coordinate axes. The cube is on the x-axis distant r_2 from the origin ($r_2 >> s$). Determine the mean intensity and the flux of the combined radiation of the two sources as measured at the origin.

2. Black body radiation of temperature T_1 passes through a box containing atoms in LTE at temperature T_2, and then it is observed. Show that a line formed by the atoms in the box will be in absorption (that is, I_ν will be decreased by the atoms as compared with surrounding frequencies) if $T_1 > T_2$, and it will be in emission if $T_2 > T_1$.

3. What determines the ratio of the intensities at two different frequencies emerging normal from a semi-infinite atmosphere if
   a) the source function is constant with depth and
      1) the absorption coefficient is constant with frequency?
      2) the absorption coefficient varies with frequency?
   b) the source function varies with depth and
      1) the absorption coefficient is constant with frequency?
      2) the absorption coefficient varies with frequency?

4. The absorption coefficient in a star is nearly constant with frequency except in a very narrow region. In this region k rises to a sharp maximum 10 times normal, then it suddenly drops to a sharp minimum 0.1 normal, then it rises again to its normal value. What does I_ν look like

in this frequency region if

a) the source function increases into the star?

b) the source function decreases into the star?

5. A uniform, plane slab has the same absorption coefficient as the star mentioned in question 4. above. What does the intensity look like in the frequency region in which k is strongly varying for various values of the thickness of the slab, that is, from the case of the slab being optically thin for all frequencies through the case in which it is thick at all frequencies?

6. In a uniform slab there are two types of processes: in the first the absorption k_1 = constant and the source function is the Planck function; in the second, k_2 has a sharp maximum at frequency ν_o and rapidly goes to zero on either side, while the source function vanishes. Sketch the variation of I_ν with frequency in the neighborhood of ν_o for various thicknesses of the slab.

7. A uniform slab has the following two absorption and emission coefficients:

$$j_{1\nu} = e_1\nu^2 \; ; \quad k_1 = a_1(\nu - 100\nu_o) \text{ if } \nu > 100\nu_o$$
$$k_1 = 0 \quad \text{if } \nu < 100\nu_o$$

$$j_{2\nu} = e_2\nu^{-2}; \quad k_2 = a_2\nu^{-3}$$

The e's, a's, and ν_o are constants. At the frequency $\nu = \nu_o$, $j_{1\nu_o} = j_{2\nu_o}$. Also the slab has an optical thickness of 100 at this same frequency ν_o Sketch the schematic variation of I_ν with frequency over the entire spectrum.

8. Prove equation (5.5).

9. A gray atmosphere in LTE has convection such that the radiative flux is

$$\pi F_r = \sigma T_e^4 \qquad \tau < \tau_o$$

$$\pi F_r = \sigma T_e^4 \frac{\tau}{\tau_o} e^{-(\tau/\tau_o - 1)} \qquad \tau > \tau_o$$

Find the temperature distribution $T(\tau)$ using the Eddington approximation. Note that T^4 approaches the finite limit $T_e^4\,(1/2 + 9\tau/4)$ as $\tau \to \infty$.

10. Outline the steps by which the structure of a model atmosphere could be found if the pressure-density relation (instead of the temperature-optical depth relation) were known.

11. Show that if the temperature is not high enough to appreciably excite helium, a model atmosphere with X, Y, and g for the H abundance, He abundance, and gravity, will be nearly identical to a model having X',

Y', and a gravity $g' = (X'/X)g$.

12. Describe how the frequency distribution of the emitted radiation will be affected if the importance of convection in a model atmosphere is changed. (The answer depends on departures from grayness.)

13. Would limb darkening $I_\nu(0,\mu)/I_\nu(0,1)$ show greater contrast as measured in continuum radiation or as measured in the center of a strong line?

14. All atoms in a gas are moving with the same speed but in random directions. Find the profile function due to the Doppler effect.

15. Supergiant stars have much turbulence in their atmospheres, but small pressures. Dwarfs have high pressures but little turbulence. Assuming this turbulence is all of the micro—variety (it is not in reality), compare the appearance of lines in a supergiant and in a dwarf of the same temperature.

16. If a small amount of hydrogen in the Sun were replaced by helium, how would an absorption line of sodium be affected? Of hydrogen? Of helium?

17. The Planck function varies with continuum optical depth as

$$B_\nu = B_0(b\tau_c + 1) + Ae^{-\alpha\tau_c}$$

with $\alpha >> 1$. This form simulates the chromospheric temperature rise. Show that the condition for the center of a very strong line to go into emission ($I_\nu > I_c$) is approximately that $A > bB_0$. Assume LTE.

18. The Planck function at a certain frequency is (in arbitrary units)

$$B_\nu = 5 \qquad \tau_c < 10^{-3}$$
$$B_\nu = 1 + \tau_c \qquad \tau_c \geq 10^{-3}$$

Find approximately the quantitative relation between I_ν/I_c and u for an LTE line that is borderline in strength between

a) weak and medium
b) medium and strong.

19. Two lines of an element have the following properties:

	f	λ	g	χ
line 1	0.03	6709 A	3	5 eV
line 2	0.09	6730 A	10	6 eV

In an early A star (T is about 10^4 °K), how does the ratio of the two equivalent widths vary with the abundance of the element?

20. A spherical star rotates with an equatorial velocity V. Show that a photon absorbed or emitted by an atom in the star at frequency ν_0

will be observed in the total flux from the star to have a frequency between ν and $\nu + d\nu$ with the probability

$$p(\nu)d\nu = \frac{2}{\pi}\left\{1 - \left[\frac{c(\nu - \nu_0)}{\nu_0 V \sin i}\right]^2\right\}^{1/2} \frac{c d\nu}{\nu_0 V \sin i}$$

for frequencies satisfying $(\nu - \nu_0)^2 \leq (\nu_0 V \sin i / c)^2$. i is the angle between the rotation axis of the star and the line of sight. Ignore complications such as other broadening mechanisms and limb darkening of the star.

INDEX